# THE DECLINE AND FALL OF THE HUMAN EMPIRE

Why Our Species Is on the Edge of Extinction

## HENRY GEE

ST. MARTIN'S PRESS
NEW YORK

First published in the United States by St. Martin's Press, an imprint of
St. Martin's Publishing Group

THE DECLINE AND FALL OF THE HUMAN EMPIRE. Copyright © 2025 by Henry Gee.
All rights reserved. Printed in the United States of America. For information, address
St. Martin's Publishing Group, 120 Broadway, New York, NY 10271.

www.stmartins.com

The Library of Congress Cataloging-in-Publication Data is available upon request.

ISBN 978-1-250-32558-7 (hardcover)
ISBN 978-1-250-32559-4 (ebook)

Our books may be purchased in bulk for promotional, educational, or business use.
Please contact your local bookseller or the Macmillan Corporate and Premium
Sales Department at 1-800-221-7945, extension 5442, or by email
at MacmillanSpecialMarkets@macmillan.com.

Originally published in Great Britain by Picador, an imprint of Pan Macmillan

First U.S. Edition: 2025

10 9 8 7 6 5 4 3 2 1

# Contents

*Prologue* 1

## PART ONE: RISE

1: The Human Family 31
2: The Genus *Homo* 44
3: Last Among Equals 57
4: Last Human Standing 70

## PART TWO: FALL

5: Agriculture: The First Casualty 85
6: Pox-Ridden, Worm-Eaten and Lousy 99
7: On the Brink 114
8: Over the Edge 137
9: Free Fall, and After 154

## PART THREE: ESCAPE

10: The Future Is Green and Female 171
11: Turning Over a New Leaf 188
12: Expanding the Human Niche 203

*Afterword* 234
*Notes* 236
*Index* 271

# PROLOGUE

*In a composition of some days, in a perusal of some hours, six hundred years have rolled away, and the duration of a life or reign is contracted to a fleeting moment: the grave is ever beside the throne; the success of a criminal is almost instantly followed by the loss of his prize; and our immortal reason survives and disdains the sixty phantoms of kings who have passed before our eyes, and faintly dwell on our remembrance.*

EDWARD GIBBON,
*The Decline and Fall of the Roman Empire*

Of all the animals that have risen and fallen in Earth's long history, it is the dinosaurs that have most engaged the public imagination. They infest our cinemas and comics. Their images are all over clothing and school lunchboxes. They even get into places that they're not wanted. The word 'dinosaur' has such cachet – such pulling power – that it can even sell things that don't have much to do with dinosaurs. *Dawn of the Dinosaurs* is the title of a wonderful book[1] about the unjustly neglected Triassic Period (the one before the Jurassic of movie fame, for all that many of the dinosaurs featured therein are from the even later Cretaceous

Period), a fifty-million-year interval (between 250 and 200 million years ago) in which many kinds of extraordinary creatures evolved, most of which despite their inherent interest will be unfamiliar to most people, and in which dinosaurs only played a walk-on part near the end. But they got top billing on the title page, nonetheless. Such palaeocultural misappropriation doesn't stop there. Popular books on prehistoric life are often entitled 'Dinosaurs', and then, in smaller letters, as if in apology, 'and Other Prehistoric Animals'. Play sets featuring dinosaur models might include animals that lived at the same time as dinosaurs but weren't dinosaurs (the flying pterosaurs, the swimming plesiosaurs) and even *Dimetrodon*, a toothy, sail-backed creature that lived tens of millions of years before the dinosaurs and was in fact more closely related to us. As if to stick two fingers up at the merest accusation of anachronism, they might even throw in a woolly mammoth, an animal that lived more than fifty million years after the dinosaurs had disappeared. Such is the near synonymy of the word 'dinosaur' with 'clickbait' that I was tempted to call this book 'Dinosaurs' (in large capital letters) with the subtitle (in smaller letters) 'and Their Role in the Evolution and Extinction of Humans'.

They are not, however, completely unconnected. Whatever else one might say about dinosaurs, they are, indubitably and most definitely, extinct.[2] The fascination that they elicit in the public mind is derived, in large measure, by this single, simple fact. Dinosaurs seemed to be thriving until the very end of the Cretaceous Period about sixty-six million years ago, when, suddenly, they disappeared, along with the flying pterosaurs (which weren't dinosaurs), the swimming plesiosaurs (and neither were these), the iconic truck-tyre-sized ammonites (armoured relatives of squid, definitely not dinosaurs) and many other creatures.

For many years – decades, even – the reason why dinosaurs perished was a non-question that attracted a wealth of possible answers. More than a hundred have cluttered up the cocktail-party-chat end of the scientific literature.[3] Dinosaurs died out because their eggshells became too thin and fragile for the embryos to reach term. Dinosaurs died out because their eggshells became too thick for the young to hatch. Dinosaurs died out because newly evolved mammals ate their eggs (irrespective of the thickness of the shells). Dinosaurs died out because of indigestion brought on by eating the then newly evolved flowering plants. Dinosaurs died out because of hay fever contracted from the then newly evolved flowering plants and sneezed to death. Dinosaurs died from some other unspecified contagion. Dinosaurs died out because they became too big and heavy. Dinosaurs died because they stopped having sex. Dinosaurs died out because, after 160 million years as the undisputed rulers of creation, they simply ran out of things to do, new worlds to conquer, and got bored (a condition that's accreted its own technical term, *Palaeoweltschmerz*), and expired from sheer *ennui*.

The most outrageous idea to account for the extinction of the dinosaurs was that the world was struck by an enormous asteroid, the impact of which ushered in an interval of global catastrophe. This was the idea that turned out to be true.

Many of the ideas, though, turned on the notion that after a hundred million years and more of tenure on the

Earth, the dinosaurs had become outmoded, tired, worn out, that their time was over. This, really, was why extinction was a kind of non-question, for it was once thought that dynasties of creatures came and went as a matter of course, and that departure was inevitable, the creatures having run out of puff. It was called 'racial senescence', or, more technically, 'orthogenesis'. And so it was that the Devonian Period was the 'Age of Fishes', after which fishes gave way to the succeeding 'Age of Amphibians' in the Carboniferous Period, with each dynasty of creature giving way to something newer, more advanced, more evolved, with each in its allotted station, to each its allotted span. The dinosaurs were the apotheosis and zenith of the 'Age of Reptiles', after which there was the 'Age of Mammals', culminating in Man. (Capital M. Not, you'll notice, Woman, capital W.) Dynasties of animals arrived, largely supplanted the existing ones, and were, in their turn, eclipsed. Dinosaurs arrived, fretted and strutted on the stage of life, and then departed when their time was up. So much so that the word 'dinosaur' became a synonym for something huge, obsolescent and unsuited to the modern world, like the Ford Edsel or manual typewriters.

Only in the 1970s were dinosaurs reinvented as warm-blooded, intelligent creatures, largely thanks to the patient work of the late palaeontologist John H. Ostrom and his swashbuckling student Robert Bakker. A little later, the idea that an asteroid brought the age of dinosaurs to a crashing close gained credence, and it was found that dinosaurs didn't die out because of some natural cycle, but that they were cut down in their prime. If the asteroid had missed, dinosaurs might still be enthroned on the Earth today.

After this, orthogenesis was dumped in the 'where are they now?' files of defunct evolutionary ideas. It was an idea that had had its day, like, well, the dinosaurs. Evolution, you see, doesn't work that way. The motor of evolution is natural selection, a handy phrase for what happens when inherited variation and superabundant offspring confront environmental change. When this collision is allowed to play out over time, the result is evolutionary change. But at any given moment, natural selection has no memory of things past, no vision for the future, no aim in view. This in-the-moment view was encapsulated by palaeontologist Leigh Van Valen as the 'Red Queen' hypothesis. Creatures are forever in competition. Predators develop keener weapons and sharper senses to catch prey that are forever evolving to be more alert, more wary. As the Red Queen says to Alice in Lewis Carroll's *Through the Looking Glass*, in Looking-Glass Land you must run as fast as possible just to stay in the same place. From this, Van Valen reasoned that there need be no relationship between how long a species (or group of species) took a turn on the stage of life and the timing or manner of its passing. Orthogenesis – racial senescence – was an illusion.

Except that it wasn't.

Time after time, palaeontologists found that species (or, more usually, genera, which are groups of closely related species) tended to arrive at one point in Earth's history, after which they diversified and became dominant, only to fade and dribble away to extinction. If orthogenesis made no sense, what could explain this pattern of entrances and exits?

An answer came as recently as 2017, when a group of

palaeontologists from Helsinki came up with a way to finesse the problem.[4] Species appear on the scene for any number of reasons, but their rise to dominance is driven by having to rub up against other creatures, in the same way that it takes a bit of irritating grit in an oyster to create a pearl. Once they've reached the top, having run all the competition out of town, they must fight a long and inevitably losing battle against a pitiless foe that will never, ever give up – the Earth itself. After that, the only way is down, until, eventually, the tiny remnant of what is left will be culled by some chance environmental event. The asteroid that mowed down the dinosaurs was a freak event. If the asteroid had missed, the dinosaurs would have died out, sooner or later, because that's what living things do. To understand when and why species become extinct in the common run of things, you must see what they were doing at their peak.

And so, we arrive at the connection between dinosaurs and humans.[5] If you apply this logic to modern humans, our own species, *Homo sapiens*, is marked for extinction. Of course it is. Humans, like dinosaurs, and like every other living thing that has ever walked, hopped or crawled on this planet's face, or has burrowed beneath, or flown above it, or slithered into the ocean depths, are living creatures and cannot claim any special exemption from the rules of the world. On the other hand, *Homo sapiens* might not be an exception, but it is exceptional, nonetheless, and I think it's fair to state this even though a human is doing the judging.[6] Humans have achieved, in a remarkably short space of time, a dominance over the fates of all other creatures that none other has managed in the entire history of the Earth, as far as is known. They have managed this with an unprecedented mastery over the forces of nature. Humans have penetrated the secrets of atoms, to exploit their energy. They have unravelled the genetic mysteries of life, to manipulate it. They have scanned to the edges of the Universe. It is reasonable to ask whether, given all these achievements, humans might, uniquely, evade extinction's great scythe, and persist indefinitely.

The answer is that humanity will perish, along with everything else. The reason, in part, is because of its very success. Humanity has become so pre-eminent, so dominant, that it threatens the functioning of the ecosystem on which it and all other creatures depend. No other single species in the history of creation has ever posed such a threat.[7]

In my previous book, *A (Very) Short History of Life on Earth*, I was coy about precisely when *Homo sapiens* might vanish. I veered between a few thousand years, to a few tens of thousands, to – well – some vague, hand-wavy time in the future. I think the phrase I used was 'sooner or later'. In a way, extinction isn't some uniform daisy-cutter, but affects each species according to its own circumstances. In that book I was proud to have invented the Karenina Principle – that whereas all happy, thriving, abundant species are the same, each species on the brink of extinction will meet oblivion in its own way.

There are, however, rules. According to the principles discovered by the Helsinki group, our species' inevitable demise was set the moment when, for the first time in the seven-million-year history of the human lineage, only one human species was left standing. That species was *Homo sapiens*. The moment? At some point between 50,000 and 25,000 years ago. After that, the downward course became inexorable. The only question was when the final curtain would fall.

Until about 50,000 years ago, *Homo sapiens* was just one of many species of human. At that time, it was tropical, living in Africa and southern Asia, with a toehold in island south-east Asia and Australia. In Europe and central and northern Asia, Neanderthals ruled, ruggedly adapted for ice age life.[8] They shared central and eastern Asia with the Denisovans,[9] initially adapted for the high life on the Tibetan Plateau, but which had lately migrated to lower altitudes. Island south-east Asia hosted a remnant of the ancient species *Homo erectus* in Java, and the unusual, dwarfed 'hobbits' *Homo luzonensis* in the Philippines and *Homo floresiensis* on the island of Flores, in what is now Indonesia. And these are just the ones we know about. *Homo sapiens* shared Africa with other species, the existence of which we are barely aware. But by around 25,000 years ago, coincident with the coldest snap of the ice age, *Homo sapiens* had colonized the whole of Africa and Eurasia and had even made inroads into the Americas. By that time, it had driven all other human species to extinction. From the moment the last human that wasn't a member of the all-conquering *Homo sapiens* died out, *Homo sapiens*' card was marked.

It then had to fight the long, losing battle against ineluctable fate.

This might seem odd given that the human career since then appears to have been one of progressive global domination. Beneath the success story, however, *Homo sapiens* has been living on borrowed time. It has done so for thousands of years and is now on the verge of collapse.

In this book I shall explain why.

I shall also explain how, if it is wise, lucky and sufficiently ingenious, *Homo sapiens* might evade this fate, at least for a while. But if it is to do this, it must act now. The reason is that humans alive now, including you, the reader,[10] live at a unique turning point in the long history of the species.

In seeking a place to begin his magisterial, 2,900-page epic *The History of the Decline and Fall of the Roman Empire*, Edward Gibbon did not start with the founding of Rome itself.[11] Neither did he discuss the wars with Hannibal of Carthage; the campaigns of Scipio, Pompey or Mark Antony; Caesar's conquest of Gaul; or the subversion of the Roman Republic by the Empire. Instead, he began when the Empire was at its height, under Trajan, at the start of the second century CE.[12] He understood, instinctively, that to chart the course of the Empire's decline he needed to begin when it was at the top – rather as the Helsinki group of researchers did with species in general.

Gibbon did, to establish context, look back slightly, just a century, to the days of Augustus, and especially to the Augustan view that the Roman Empire, even then, had reached its natural limits – the Mediterranean basin, bounded by the Rhine, the Danube, the Euphrates and the Sahara – and to exceed these bounds would be counterproductive. So it proved. Although the legions did, at times, march across the Rhine into Germany, and across the Danube into what is now Romania, these conquests were short-lived. At times, the sounds of Roman legionaries on the march were to be heard on the coasts of the Caspian

Sea and the Gulf. But campaigns in such faraway places were fleeting, evanescent and soon thrown into reverse. The Augustan boundaries were to prove enduring.

As Gibbon measured the decline of Rome from its period of greatest glory, I shall start my charting of the decline of the human species a little before the time when *Homo sapiens* found itself at the top of the heap, the sole survivor of the hominin (human) family.[13] In fact, I shall go a bit further back than that, by beginning with the earliest hominins. The reason is that our family – the family of the hominins – is defined by a single fact, that is, a habitually bipedal stance. Hominins are those apes that got up on their hind legs and walked. Many of the things that subsequently happened in the human lineage, from the evolution of the overlarge brain, the long childhood, the invention of technology and so on, might not have happened had we stayed with all four feet on the ground.[14] But bipedality, for all its apparent virtues, is thoroughly bad news for human health and welfare. It required a comprehensive re-engineering of a body which, for the previous 500 million years, had evolved with a backbone held horizontally, in tension. The sudden conversion to a backbone held vertically and in compression prompted a plethora of problems, which even today remain a challenge. From hernias to piles, back pain to broken bones, walking on two legs continues to exact a fearsome toll.

I shall then focus on the origins of just one genus of hominin, the genus *Homo* – that particular group of species, nested within the hominins, that includes our own. Until the arrival of *Homo*, between three and two million years ago, all hominins were restricted to Africa. With *Homo*, though, they left their ancestral homeland and spread across the world, evolving into a host of new species from the hobbits of south-east Asia to troglodyte Neanderthals, to the heavily built Denisovans that colonized Siberia and Tibet. All, though, were rare and scattered. While early members of the human family were omnivorous opportunists, they were never more than marginal players in the ecosystem. Even when some became specialist vegetarians, no herds of humans roamed grazing across the plains, to match, say, zebras or wildebeest. The most successful early species of *Homo* was *Homo erectus*, which became a savannah pack-hunting predator, rather like the modern African hunting dog – and the economics of ecosystems mean that predators are always rare.

But whether hunter or gatherer, scavenger or predator, the fossils of hominins – the material remains that yet survive from millions of years past – are notoriously rare. Compared with the fossils of the other animals with which they shared their environment – from antelopes to elephants – hominin fossils are vanishingly scarce. It is likely that hominins always were rare in life, peripheral members of any ecosystem they inhabited, never present in any great numbers. When *Homo erectus* evolved to become a specialist meat eater, this scarcity became more pronounced still. Scarcity has always posed problems for hominins, as it limits the amount of genetic diversity available in a population at any given time. It also gives free rein for undesirable mutations to spread, which would be weeded out of a population were it larger (I shall explain why that is so later in the book). More seriously, scarcity means that a population is that much closer to being wiped out altogether, not by any inherent genetic flaw, but by idiosyncratic circumstance and unhappy chance – the Karenina Principle in action. It seems to have been a feature of hominins that their populations were cut down to almost nothing not just once, but perhaps several times in their history. Recent work has shed light on a particularly awkward age, during which time humans teetered on the edge of oblivion for more than 100,000 years, between about 930,000 and 813,000 years ago. During this interval the entire stock of breeding humans was never more than 1,280 individuals.[15] A modern conservationist transported back in time might have reckoned that humans were a species threatened with extinction and put it on the Red List.

Although there are more humans on Earth now than ever before, this early scarcity has left an imprint in our genetics. For all our superficial variety, we are much the same underneath. There is more genetic variation in a troupe of chimpanzees in Africa than in the entire human species.[16] This is a sign that *Homo sapiens* has expanded from very small groups of founders that managed to dodge extinction sometime in the past – probably more than once.

*Homo erectus* and its descendants moved and evolved across Eurasia, but *Homo* continued to evolve in Africa, too. In Chapter 3 I record the origin of our own species, *Homo sapiens*, in Africa sometime around 315,000 years ago, at about the same time that Neanderthals were emerging in Europe. At that time our species was something of a raw ingredient that had to be winnowed by experience. Although *Homo sapiens* tried on occasion to leave Africa, all its attempts failed until around 100,000 years ago. By then, *Homo sapiens* had weathered a long period of splitting up into smaller populations and then joining up again, interbreeding within Africa[17] to produce a creature we'd recognize today – a creature with self-knowledge and an appetite for destruction.

But *Homo sapiens* very nearly didn't happen. For most of its existence, it was bottled up in its African homeland by its cousins, the Neanderthals, which had come to dominate Europe and western Asia. As the global climate continued to cool and become drier, *Homo sapiens* almost became extinct. The last, scattered remnants of *Homo sapiens* did, however, survive – but the close encounter with oblivion left its genetic mark. Because all modern humans descend from the tiny population of ancestors that survived, the genetic resources available to humans – to overcome new challenges such as disease – are scant.

Starting roughly around 100,000 years ago, *Homo sapiens* managed to break out of Africa. This time, migration was a raging success. Humans reached Australia more than 60,000 years ago and Europe 45,000 years ago. Wherever *Homo sapiens* went, it left destruction in its wake. Like no other hominin species, it started to modify the landscape to suit its needs. The result was the extinction of most animals more massive than a large dog.[18] By around 25,000 years ago (at the latest), *Homo sapiens* had occupied all the major continental land masses. Only New Zealand, Madagascar, the more remote oceanic islands and Antarctica remained unvisited, and even they would succumb. It's worth saying that the spread of *Homo sapiens* has been so rapid, against the scale of events in geological time, and even when set against the overall hominin career, that the sudden invasion of *Homo sapiens* might be said to encompass the Moon, and – in the form of its technological tentacles – the entire Solar

System. Radio and TV signals emitted by *Homo sapiens* have spread more than a hundred light years into the wider Galaxy, encompassing a volume that includes thousands of stars.

As well as contributing to the extinction of most of the Earth's large animals, *Homo sapiens* drove all other human species to extinction – Neanderthals, rulers of Europe and Asia for more than a quarter of a million years, crumpled before the surge of modern humans entering Europe around 45,000 years ago. Although long resistant to *Homo sapiens*, the Neanderthal edifice was washed away like a sandcastle before the tidal surge. It took less than 10,000 years. At around the same time, the arrival of humans spelled doom for the yeti-like Denisovans of Asia, the hobbit-like indigenes of island south-east Asia (*Homo luzonensis* and *Homo floresiensis*) and presumably others that remain to be discovered.

At this point, the only way was down.

Evolutionary theory suggests that species succeed when they have sparring partners. As the Helsinki group showed, when the competition is eliminated, stagnation sets in and a species becomes more subject to circumstances of the external environment, as well as forces acting from within. As for Gibbon's Roman Empire, so with the human species. It is at this point that I start to document the decline of *Homo sapiens*.

The rot set in with a remarkable human innovation that is almost (but not quite) unique in the animal kingdom. That innovation is domestication of animals and plants. Starting around 40,000 years ago, when humans domesticated dogs, they tamed various other large animals as sources of meat, milk and fibres. Ferocious wild oxen and boar became Buttercup the Cow and Porky the Pig. Skittish sheep became fluffy and docile. Most of all, humans started to tame wild plants. This innovation – agriculture – started around 26,000 years ago, when the Earth was at the peak of the last ice age, and when many parts of the world not covered in ice were harsh and dry. Hunters and gatherers kept starvation at bay by farming predictable crops rather than relying on increasingly unpredictable forage. It was then that the human population started to expand exponentially – at ruinous cost. The consequences of agriculture can be felt today in a range of diseases and health problems that afflict humans, from tuberculosis to parasitic infestations to diabetes.

Perhaps more worrying is that most humans are fed from a limited selection of genetically similar crop plants. History shows that overdependence on a single crop can lead to disaster. One thinks of the Irish Potato Famine in the nineteenth century, in which the potato crop was ruined by a fungal disease. The consequence was a marked reduction in the population of Ireland, by both starvation and emigration. Today, plant breeders must keep one step ahead of the fungi that destroy the very small range of commercial varieties of banana.[19] The world could, probably, survive without bananas – but more serious would be some pathogen that afflicted wheat, rice, barley, millet or sorghum, the plants on which billions depend as staple food crops.

Human beings, like bananas, are extraordinarily susceptible to disease,[20] compared with, say, chimpanzees. The roots of this vulnerability lie deep in the past, during those anxious periods when the population of humans declined to tiny numbers, only to be re-established, resulting in a relatively small degree of genetic variation when the population did recover. This has left humans vulnerable to diseases, as attested by the history of diseases from plagues in ancient times to COVID-19 today. This phenomenon is most apparent in the many cases of small populations prone to certain disorders, whether through infection, or, as it may be, inbreeding, including the incidence of porphyria among Afrikaners,[21] bipolar disorder in the Amish[22] or Crohn's disease in Ashkenazi Jews.[23] Indeed, despite improvements in our medicine, hygiene and lifestyle, the burden and threat of disease are great. Compared with humans, chimpanzees suffer far fewer diseases, despite being strangers to personal hygiene and fond of snacking on their own faeces. Indeed, some of the diseases that afflict chimps may have been contracted from humans. One could even argue that the human burden of disease is magnified *because* of modern medical interventions, which allow many people to survive who might have perished had they been unfortunate enough to have been born in earlier centuries.[24]

A significant milestone for *Homo sapiens* came in 2022. That was the year when more people died than were born in China, the world's most populous country,[25] for the first time since Mao Zedong's Great Leap Forward in the 1960s led to widespread famine.

The news is remarkable because of the otherwise inexorable growth in the human population since the invention of agriculture. So much so that in 1968, Paul Ehrlich wrote *The Population Bomb* about the dangers of overpopulation.[26] By coincidence, the 1960s were when the human population was growing at its fastest, at a little over 2 per cent a year. Since then, the population has done something entirely new – the rate of population increase has started to fall. To be sure, the population is still increasing, but at a slower rate overall, a shade under 1 per cent. It is still increasing rapidly where the population is still young, but in many countries – perhaps most – the fertility rate is now lower than the natural rate of population replacement.[27] The most notorious case is Japan, but it is also true in countries such as Spain, Italy, Thailand and many others.

That the human population is set for contraction is remarkable enough. But why is it happening now?

There are more than twice as many people on Earth as there were when *The Population Bomb* was written, but they are (in general terms) living longer, healthier lives than they did then.[28] This is a result of innovations in agriculture, technology and healthcare provision and – most of all – the empowerment of women. In my view, the emancipation of women is the single biggest determinant of the good health and welfare of human societies. People have better control of their reproductive choices than they once did. The concern is that their choice seems to be to delay having children, or not having any at all. It could be that people are postponing making decisions about reproduction because of a general unease about the future, either based on concern about the environment generally, or because they judge that their own economic prospects do not allow for starting a family.

Underlying this seemingly non-specific unease could be the overexploitation of resources by *Homo sapiens*. As most of the Earth's resources have been used up, there is less to go round. This might also explain, in part, why the global economy, beneath the various ups and downs, has been stagnant for twenty years. That this long-term stagnation coincides with a decline in the fertility rate might not be a coincidence. And, in the modern world, bad news travels fast. Increasing interconnectedness in financial services and travel means that financial and health shocks can travel more quickly and have more widely felt consequences than they once did. The generation that was born in the 1960s (when the tide in human population began to turn) might be the last in modern times that is guaranteed to be wealthier than their parents. The children of these children simply often cannot afford to have children of their own. And there is a deeper malaise – even if people do want to have children, they are leaving that choice for later and later, increasing the risk of running into fertility problems. To cap it all, human sperm count has dropped markedly in the twentieth century, for reasons nobody can quite fathom.

The immediate consequence of population decline will be political and social upheaval. Climate change is making some parts of the Earth uninhabitable for human beings and, crucially, the crops on which they depend, particularly in the Global South. Its younger residents, unable to survive at home, are moving to the Global North. If countries in the Global North are to remain viable, they should be looking for ways to accommodate this influx as an asset, not turn it away. Most people on the move in the coming century will be from Africa, where the human population is youngest and still experiencing rapid growth.

Humans have migrated from Africa twice before. First, when *Homo erectus* became the first hominin to leave that continent around two million years ago. Second, when *Homo sapiens* migrated from Africa, starting around 100,000 years ago, eventually displacing all other hominins in Asia and Europe. Neither of these two migrations had any specific intent, or destination in mind. The people concerned migrated in response to population pressure, or climate change, or because they were following herds of game animals. Furthermore, these migrations took many generations to happen. Any single individual might have moved only a few dozen miles from their birthplace during their lifetime. What we see today as a pattern of migration is constructed in retrospect. The third migration will be quite different. It will be entirely purposeful and take a few decades at most. Any one individual will accomplish the entire journey in a space of days – months at most – and there will be more of them than in the first two migrations put together.

Everywhere, conflicts will flare which, at root, have their causes in climate change and the resulting diminution of resources, especially land for farming, and water to irrigate it. One might think that fewer people would lessen the demand on resources, but the population is still rising, and will do so until the 2060s. And yet the crises over resources are happening right now, in the 2020s. The worry is that the decline in *Homo sapiens* will be accompanied (and hastened) by famine, drought, warfare and general economic hardship. Governments in the Global North would benefit from accommodating surplus populations from a Global South increasingly suffering the double blow of a population surge and climate-change-induced damage. This will help fill worker shortages left by their own falling birth rates but could run the risk of exacerbating tensions at home. Eventually, though, populations will be falling globally, posing a challenge for governments whose constitutions are, for historical reasons, predicated on increased economic growth[29] and expanding populations. There will come a time when countries in the Global North will be vying with one another to attract the most immigrants, rather than trying to turn them away.

Looking further ahead, the human population will have declined to such an extent that it will have become functionally extinct – Dead People Walking. That is, *Homo sapiens* will still be there, but in such small numbers and scattered populations that extinction by random, chance events becomes highly likely. Neanderthal Man, the closest extinct relative of *Homo sapiens*, offers a poignant case history. This species was never common and spread itself thinly across its enormous range in Europe and western Asia. In the end, the populations became so small and scattered that Neanderthals could not find mates.[30] They were faced with just two solutions – either inbreeding or interbreeding with burgeoning *Homo sapiens*. Inbreeding is a sure-fire route to extinction. The alternative option of interbreeding resulted in a situation in which Neanderthals became extinct as a distinct, separate group. In a sense, though, Neanderthals live on – in *Homo sapiens*. Every person on the planet today whose ancestry is not exclusively African carries a tiny amount of Neanderthal DNA. The difference between the Neanderthal case and that of *Homo sapiens* is that the latter cannot rescue itself by burying its heritage in the more fecund DNA of another species – because no such species now exists. As Jared Diamond once wrote in another context, there is nothing that kills a population so much as lack of people.[31]

There is another solution, however, this time suggested by the example of *Homo erectus*, ultimate progenitor of both Neanderthals and *Homo sapiens*. Having spread from its African homeland across Eurasia, *Homo erectus* evolved into a multitude of new species, each one adapted to a particular niche. Neanderthals – as descendants of *Homo erectus* – became adapted to life in the chilly north and learned to make their homes deep within caves. Other descendants became adapted to life in rainforests, a surprisingly difficult environment to conquer. Such diversification became possible only once populations of *Homo erectus* became so widely scattered that individuals and communities rarely had the chance to meet and interbreed. So separated, each charted its own evolutionary course.

Could such evolutionary diversification be the saving of *Homo sapiens*? In this book I argue that the answer is 'yes' – but as *Homo sapiens* has effectively become a single population encompassing the whole Earth, such diversification will only be possible should human beings decide to colonize space, with separate populations living in isolated habitats, or on the surfaces of the Moon, Mars and other bodies. *Homo sapiens* could live on, but might diversify into a range of new species, as different from each other as, say, *Homo erectus* is from Neanderthals. Colonizing space on a large scale is easier said than done. It will require technologies that are either in their infancy or which do not yet exist. But if history bears witness to only one thing, it is to the seemingly endless ingenuity of life in general – and of *Homo sapiens* in particular – for getting itself out of trouble.

In *The Population Bomb*, Ehrlich sounded an urgent warning that the world would soon be unable to support its human cargo.[32] That catastrophe plainly has not happened. How has *Homo sapiens* achieved this seemingly impossible, Houdini-like feat? The answers are both technological and social.

The technological part concerns the so-called Green Revolution, in which, in response to the threat of global famine, scientists in the US, Mexico and the Philippines created new, high-yielding strains of wheat and rice that would use croplands more intensively and productively than the varieties that farmers had been accustomed to growing. This began in the 1960s (and, indeed, Ehrlich refers to it as a possible solution in *The Population Bomb*). Another revolution happened about twenty years later. The 'Gene Revolution' of the 1980s allowed for the creation of crops genetically engineered to resist pests, or grow in less favourable soils, that promise to improve agricultural production even further.

But these benefits have significant costs, and the rise in population that the Green Revolution has allowed cannot be sustained indefinitely. Eventually, a ceiling will be reached when the costs involved in improving agriculture will outweigh the benefits. One could argue that *Homo sapiens* has met or even overtopped such limits, running the risk of severe depletion of resources and degradation of the environment.[33] In a way, we are back to where we were when Ehrlich wrote his visionary tract. Today, *Homo sapiens* – just one species among millions – sequesters about one third of all the products that plants create using photosynthesis, and on which almost all life depends.[34] This is plainly not sustainable. The Green Revolution, designed to avert massive famine in the 1960s, might have only postponed disaster, rather than abolishing it.

There are, however, ways for *Homo sapiens* to reduce its overdependence on plants. One, paradoxically, is to eat more plants. The reason is that the animals raised for milk and meat – cattle, goats, pigs and sheep – also eat plants, but much of the energy so gained is wasted before their meat reaches the plate. If humans ate plants directly, rather than eating animals that eat plants, plants would be used more efficiently. The end result could be that humans use less of the Earth's natural bounty, which would relieve some strain on the planet's biological diversity.

There may be even more radical solutions. One might be to find a way to subvert photosynthesis by genetically engineering organisms, such as bacteria, to make the process more efficient, a kind of second-stage Green Revolution. Another might be to convert waste products into food, either by chemical means or, more likely, with the aid of bacteria, or fungi such as yeasts. A third might be to create an entirely technological version of photosynthesis, converting carbon dioxide and water into sugars directly, without plants having to be involved at all.[35]

A side effect of all these ideas – closing ecological loops, artificial photosynthesis and so on – prompted by the need for a large population of *Homo sapiens* to live more sustainably on a finite Earth, might be that the species will acquire the technological means to live beyond it. *Homo sapiens* might be able to outrun fate by expanding beyond the planet of its birth. In this book I hope to show that going into space, aided by artificial versions of photosynthesis and other technological advances, might offer humans a way out of the dilemmas they now face, and could even delay the extinction of the species. Until, that is, the remorseless gears of evolution create a range of post-human species.

There is a sense in which *Homo sapiens* was always meant to venture into space. It's a part of its heritage to challenge environmental constraints. The roots of human technological ingenuity run deep.

Primates are, and always were, tropical animals. Primates that naturally live outside the tropics are rare and remarkable. One is the Barbary ape (*Macaca sylvanus*) – not an ape at all, but a monkey – that is naturalized in Gibraltar, in southern Spain. Another is its cousin the snow monkey (*Macaca fuscata*), which lives in chilly mountainous areas of Japan.

The genus *Homo* has taken the extra-tropical adventures of primates to another level. An early form of *Homo*, perhaps *Homo antecessor*, made its home in England more than 800,000 years ago, when the climate was temperate and deteriorating into a cold snap.[36] *Homo erectus* made fire in caves near what is now Beijing, in the north of China, another challenging environment well outside the tropics. Neanderthals ventured into Arctic Russia, and *Homo sapiens* soon colonized the rest of the world, no matter how inclement the climate.

Humans could only do this, however, thanks to technological refinements such as fire, artificial shelters and clothing. That people now live from the tropics to the poles, therefore, disguises the fact that the range of environments in which people can live without the aid of technology is narrower than one might imagine.

This is where climate change enters the picture. Even the tropical heartlands that are home to billions of people will soon become too hot and humid for humans to inhabit without further technological assistance. *Homo sapiens* is also fond of living in low-lying areas near coasts, all of which are now vulnerable to rising sea levels. Humans, as ever, will have to resort to technological solutions such that the biogeographic range of the species does not contract so much that extinction will inevitably follow.

It's no surprise that humans have already shown great ingenuity in doing this. Much of the Netherlands is below sea level and has been for a very long time, yet a great deal of that country's land has been reclaimed from the sea. The same can be said of those parts of eastern England surrounding the gulf known as the Wash. In medieval times, this area was largely underwater. Now, it is the breadbasket of England. Although rising sea levels threaten it again, it cannot be beyond the ability of modern humans to maintain and even extend the amount of productive land surface. *Homo sapiens* is learning to survive otherwise insupportably hot and humid environments by clever urban design, and inventions such as air conditioning. In the context of human evolution, space is just another seemingly hostile extra-tropical environment ripe for *Homo sapiens* to learn to inhabit.[37]

There is a catch, though. The launch window is narrow. Technological advancement requires a substantial resource base in the form of human brains. Fewer brains mean technological stagnation. The global population of humans will continue to rise until the third quarter of this century, after which it will stabilize and start to fall, perhaps precipitously. If people do not colonize space substantially in the next century or two, it might not happen at all.

Humanity therefore faces a choice. It is a choice that must be made now, at a time when *Homo sapiens* faces a series of political, social, biological and environmental crises unique in its evolutionary history, and at the inflection point when – for the first time – the population of the species is starting to shrink.

If *Homo sapiens* carries on much as it is, it will become extinct. The last in a lineage of hominins stretching back almost seven million years will vanish from the Earth, leaving no descendants. To be sure, as I have discussed, all species become extinct eventually. It's the natural and normal thing for species to do. However, invoking the Karenina Principle, *Homo sapiens* will meet its doom in its own way, the precise details of which cannot be foreseen. Environmental degradation? Nuclear or biological warfare? Global famine? Another pandemic? Artificial intelligence? A plague of killer robots? Zombie apocalypse? A mass attack of *Weltschmerz*? Whatever the cause, I shall argue that the extinction of *Homo sapiens* will come relatively soon, in geological terms – within the next 10,000 years.

On the other hand, if *Homo sapiens* migrates into space,

it has the potential to flower and flourish perhaps for many millions of years, evolving and diverging in ways that cannot yet be guessed.

The crux is that despite the long history of our species, this choice must be made now. That is, within the lifetimes of people now living.

When I was at primary school in the mid-1960s, when excitement about space travel was at its height, my school library displayed a book entitled *You Will Go to the Moon*. My infant excitement wore off. I never got to the Moon, and none of my classmates did, either. In fact, only twelve people have set foot on the Moon, and none at all for more than half a century. Titles such as *You Will Go to the Moon* seem dated, quaint even. Now, though, the sense of it, at least, seems to have acquired a new urgency. You *Will* Go to the Moon. Because, if you don't, there might be no one left on the Earth for you to wave back to.

# PART ONE
# RISE

# I

# THE HUMAN FAMILY

*We imperceptibly advance from youth to age without observing the gradual, but incessant, change in human affairs; and even in our larger experience of history, the imagination is accustomed, by a perpetual series of causes and effects, to unite the most distant revolutions. But if the interval between two memorable eras could be instantly annihilated; if it were possible, after a momentary slumber of two hundred years, to display the new world to the eyes of a spectator who still retained a lively and recent impression of the old, his surprise and his reflections would furnish the pleasing subject of a philosophical romance.*

EDWARD GIBBON,
*The Decline and Fall of the Roman Empire*

We owe so much to our mentors. I am lucky to have had many. One of them was the late Professor Robert McNeill Alexander – Neill to his friends – who was Professor of Zoology at the University of Leeds when I was an undergraduate there between 1981 and 1984. Neill was a kindly man with a long white beard who (to this impressionable student) resembled Obi-Wan Kenobi or Gandalf. He was an expert in the science of how animals move. In his laboratory he measured how much force kangaroos use when they jump, and how much energy camels store in the tendons of their legs when they walk.

He had a breathlessly simple way of working, so much so that *The Mail on Sunday* newspaper once dubbed him one of 'Britain's Nuttiest Professors – Ten Sages Who Really Know Their Onions' – of his many accolades, the one whence he derived most amusement.[1] He would estimate dinosaur mass by measuring how much water was displaced by a scale model of a dinosaur when suspended in a beaker of water. And he worked out a nifty way of estimating how fast animals move based on the spacing between their footprints. He did this by watching his children run around on a beach. He could time how fast they ran, and measure the spaces between the prints they left on the sand. He used these data with a few simple calculations to estimate an animal's walking speed from its footprints, even if the track maker was long gone – or even extinct. In such a simple way he could bring the past vividly to life. By measuring the spaces between their footprints, he reasoned that dinosaurs moved rather slowly.[2]

So, it seemed, did *Australopithecus afarensis*, an early hominin that left its footprints impressed in wet volcanic ash more than three million years ago at a place called Laetoli in what is now Tanzania.[3] Amid a riot of other animal footprints (the area was as busy as a railway station concourse at rush hour), a family of *A. afarensis* made its way from here to there at a leisurely amble. *Australopithecus afarensis* was a very different creature from modern humans. She was diminutive, with relatively short legs, and had a small brain. In many ways she would have resembled a chimpanzee. But she did one thing – just one – that was distinctively human. That is, she walked upright, as of habit. Her footprints show that she walked, on her two hind feet, almost as well as people today.[4]

She was, however, not the only biped around. Another set of prints at Laetoli was made by a biped, but one that walked in a strange way, different from *A. afarensis* or modern humans. For a long time, it was thought that the prints might have been made by a bear – an animal known to walk on its hind legs occasionally – but more recent work suggests that these, too, were made by an early hominin, perhaps a cousin of *A. afarensis*.[5] If so, all trace of this species has vanished entirely from the Earth. Unlike *A. afarensis*, which is known from bones and teeth, this second hominin left nothing more to demonstrate its existence than a trail of puzzling prints.

They are not the only enigma. One day, around three million years before the family of *A. afarensis* picked their way across the ashfall in Tanzania, something left a series of footprints in wet sand in what is now the Mediterranean island of Crete. At the time, Crete was joined by a land bridge with Greece to the north. The prints are astonishingly human-like. The track maker had five toes on each foot. The big toe was larger than the other toes, but closely pressed and lined up with them. This is important – in apes, such as chimpanzees, and, indeed, in some early hominins, the big toe is set well apart from the other toes and is more like a thumb. The heel, though, is much less well-defined than in a modern human foot. The creature would have been a biped, but, like the not-quite–*A. afarensis* track maker at Laetoli, not quite human, either. The identity of the Cretan track maker is a mystery. It might have been a fossil ape called *Graecopithecus*, which lived in the Balkans about seven million years ago. It is known only from fossils of its skull and teeth. But no leg or foot bones of *Graecopithecus* are available that we can match with the prints, like Cinderella's missing shoe. It might have been something else. Nobody knows.[6]

Modern apes – chimps, gorillas and so on – can walk upright, but they only do this occasionally and soon revert to their habitual posture of all four limbs, hands and feet, on the ground. What does seem clear, though, is that between ten and six million years ago, some kinds of ape were beginning to walk upright as of habit. Between nine and seven million years ago and somewhat to the west of Crete, in an archipelago that would one day become northern Italy, lived an ape called *Oreopithecus*. This, too, might have walked upright, though its feet were chimp-like, with divergent big toes. Another ape was *Danuvius*, which lived in Bavaria between twelve and eleven million years ago.[7] This was probably not a biped. However, it was well adapted to clambering around on tree trunks and branches in what would have looked like an upright posture. Walking on two legs would have been like moving around on branches, only without the branches.

Apes are few in today's world. Chimpanzees and gorillas live in West and Central Africa. Orangutans and gibbons live in Southeast Asia. All live in fast-diminishing tropical forests. Ten million years ago, Eurasia and Africa were home to a much larger variety of apes than is found today – but then, there was more forest. *Graecopithecus*, *Oreopithecus* and *Danuvius* were joined by dozens of others.

Almost all of them are known from teeth, sometimes

preserved in fragments of jaws. Tooth enamel is the hardest substance made by any living thing. Because of this, teeth are the parts of the skeleton most resistant to the ravages of time. Skulls and limb bones are much more rarely found as fossils, even as fragments. Whole skeletons are prizes indeed. After ten million years ago, even these shattered remnants of apes almost vanish from the fossil record, to be replaced by fossils of monkeys. The only fossil of a chimpanzee is a tooth, dated to half a million years ago.[8] There are no fossils known of gorillas or orangutans – although the African *Chororapithecus* (which lived ten million years ago) might have been a relative of the gorilla,[9] and *Sivapithecus* (twelve million years ago) and *Khoratpithecus* (about eight million years ago)[10] relatives of the orangutan.[11] The reason ape fossils are so few after about ten million years ago is that apes retreated to tropical forests, where fossilization is extremely infrequent – and because the forests themselves were shrinking as a result of the long, slow period of global cooling and drying that culminated in the ice ages from around 2.5 million years ago.

The record of fossil apes from between ten and five million years ago barely registered until, one day, a spectacular fossil dropped right into the middle of the period. That fossil was a hominin.

*Sahelanthropus* comes from perhaps the most unlikely place imaginable – a region of central Chad once lush and green but now so dry, windy and desolate that it looks more like the surface of the Moon than anywhere on this planet.[12] Researchers who return from fieldwork there look like they have been sand-blasted. Yet, even there, human beings cling on to life. The skull of what came to be scientifically known as *Sahelanthropus* was nicknamed 'Toumaï', which means 'hope of life' in the local language. The skull that was unearthed is almost complete (if a little bit squashed), quite the contrast to the usual fossils of apes or hominins, which are either chips of tooth, or bones that look like cornflakes that have been run down by a steamroller. Toumaï is between seven and six million years old – and was a biped.

You might be excused for wondering how you can tell whether an animal was a biped when all you have in your hand is a fossil skull, with no trace of arms or legs. The reason lies in the hole at the base of the skull where the backbone joins and the spinal cord passes through to link up with the brain. This hole is called the foramen magnum (which is just Latin for 'big hole'). In quadrupeds – animals that keep all four feet on the ground, such as dogs, or horses, with the backbone held horizontally – the foramen magnum is at the back of the skull. In hominins though, which are bipeds, with the backbone held vertically, it is tucked underneath the skull. This ensures that hominins still face forwards, even though their backbones have effectively been turned through a right angle. The skull of *Sahelanthropus* looks similar to that of a chimpanzee, except that the foramen magnum was placed more towards the base of the skull than is the case in chimps, suggesting that, in life, *Sahelanthropus* was more of a habitual biped. Part of a thigh bone and some pieces of elbow, attributed to *Sahelanthropus*, have since been described. They suggest that it was a biped, though it retained quite a few adaptations for climbing.[13]

The barely visible record of hominins picks up again with *Orrorin*, which lived in Kenya about six million years ago and is known from a few fragments, which – crucially – include part of a thigh bone, showing that it was a biped.[14] Other early bipeds were *Ardipithecus kadabba*, from Ethiopia, which lived five million years ago,[15] and a later relative, *Ardipithecus ramidus*, which lived 4.4 million years ago in the same region.[16] Curiously, though, a skeleton of *Ardipithecus ramidus* showed that its commitment to bipedalism was equivocal, and it would probably have been found in the trees as much as on the ground. An isolated skeleton of a hominin foot discovered in Ethiopia and dated to 3.4 million years ago has a mobile big toe, more like a thumb;[17] this is about the same age as the footprints from Laetoli. It confirms that more than one species of hominin was living in Africa at the time, and that some were more committed to life on the ground than others.

The walkers won out over the climbers, however, as the forests began to wither away, leaving a country of open savannah only punctuated here and there by isolated patches of woodland. All hominins after *Australopithecus afarensis* were committed bipeds. Even in *A. afarensis*, climbing ability appears to have been waning. Forensic analysis of the partial skeleton of the *Australopithecus afarensis* known as 'Lucy' shows that she likely died from injuries sustained after falling out of a tree.[18]

The big question is *why*? Why did hominins adopt bipedality as a matter of course, rather than just now and then? The answer is (spoiler alert!) that nobody knows. Questions of 'why' are always the hardest to answer. However, the unusual nature of this arrangement is highlighted when one realizes that no other mammals are habitual bipeds.[19]

The conventional explanations – or at least the ones that tend to have the most public currency – would have it that bipedality was all to the good. And there is no shortage of ideas. Hominins stood upright so they could free their hands for holding babies, making tools or carrying food, or combinations of all three. Hominins stood upright to make it easier for them to peer over tall grass in open country. Hominins stood upright to make it easier for them to wade in deep water.[20] Hominins stood upright in order that they could show off their more tender parts to prospective mates. The problem is that all such explanations are made after the fact. Certainly, walking on two legs allows for all these things, and more, but such scenarios do not account for how upright walking happened in the first place (still less, why). Indeed, two things seem to militate against it happening at all.

First, many animals carry babies, make tools, carry food, move around in water and so on quite well while remaining entirely four-footed. Chimpanzees, and even some monkeys, make tools – perhaps even of a sophistication equivalent to those made by the earliest tool-making hominins – without the need to free their hands on a permanent basis. So much so that 'primate archaeology' – the excavation of tools made by monkeys and apes in the past – has become a new specialism.[21] Many monkeys, such as baboons, live in open country and remain resolutely quadrupedal.

Second, standing upright isn't easy. Those mammals that stand upright for short lengths of time (apes, bears, meerkats, performing dogs and so on) find it tiring, which is why they soon revert to a four-footed posture. To be bipedal by habit requires a complete re-engineering of the body. It starts with the backbone.

When the backbone first evolved in the earliest ancestors of fishes, more than 500 million years ago, it was as a horizontal beam that anchored and provided purchase for the muscles of the body wall and served as a hanger from which to suspend the other internal organs. The presence of a backbone defines the vertebrates – the great group of animals to which we belong. In nearly all vertebrates, the backbone is horizontal. This is even the case in the many bipedal reptiles, including dinosaurs, and – of course – birds. Dinosaurs could become bipeds by virtue of a long tail, which acted as a counterweight to the horizontally poised part of the body forward of the hips. Birds have short tails, but they keep their backbone horizontal by maintaining an unusual, crouched posture, with the hind limbs permanently flexed at the knee.

In adopting an upright stance, hominins have turned the backbone through 90 degrees – from a horizontal beam held in tension to a vertical pole held in compression. That this is a wholly radical transformation cannot be stressed enough. The question, then, is not so much why hominins became bipeds, but how it was that they had painted themselves into such an evolutionary corner that no other option was possible.

One possible reason was the lack of a tail. Apes, unlike monkeys, do not have tails. Most monkeys have long tails that can be used as counterweights while climbing and balancing, rather in the same way that tightrope walkers use their long poles. In the so-called New World monkeys of the tropical Americas (which are only distantly related to the Old World monkeys and apes of Eurasia and Africa), the tail has become a fifth limb, capable of gripping branches independently. Admittedly, some monkeys, such as baboons and macaques, only have short tails, but these species have adopted a life on the ground – where they remain steadfastly four-footed.

Could ground life be a reason why apes lost their tails? The answer is 'no', because apes, even without tails, are still creatures of forests, spending a good deal of their lives in the trees, and have found ways to get around without tails. Orangutans use all four limbs as hands, allowing them to clamber among branches. Gibbons suspend themselves beneath the branches, swinging between them using their long arms. As well as being tailless, apes are rather larger than monkeys, as a rule, and this increase in size could be key to the adoption of bipedality by hominins. Too large to run along branches and jump between them, apes must either suspend themselves from them and swing beneath them, or clamber around among them. It could be that lifestyles such as these tend to offer a chink of advantage to an animal that can adopt an upright posture as well as a horizontal one. As I discussed earlier, *Danuvius* – that ape that lived between twelve and eleven million years ago in what is now Bavaria – seems to have been built to clamber among branches rather than walk along them, a lifestyle that presupposed bipedality. Living in a world where trees were getting scarce; already tailless by virtue of their ape ancestry, and so unable to use their tails as counterbalances; and having grown to sizes larger than monkeys – hominins had no option but to get up on their hind legs and walk.

Bipedality is inherently dangerous. As anyone who's watched a three-legged dog run around will attest, at least some quadrupeds can manage to walk, even run, after the loss of a single limb. For a biped, though, the loss or even temporary injury of a limb renders the animal almost completely immobile, and thus at higher risk of exposure to predators. Once hominins had become bipeds, then, they had to get very good at it, very quickly. As I discussed earlier, the footprints of *Australopithecus afarensis* show that hominins were walking in an essentially modern fashion more than three million years ago. Within a few million years, hominins had re-engineered their bodies to be vertical rather than horizontal.

In the head, the repositioning of the foramen magnum was just the start. *Homo sapiens* also has a ligament that attaches the back of the skull to the backbone, making sure the head doesn't droop. The spine has a distinct curvature not found in quadrupeds: It curves slightly backwards at the thorax, forwards in the abdominal region and markedly backwards again where it joins the pelvis. The pelvis itself becomes shorter and wider, with pronounced flares for attaching the huge gluteus maximus muscles – the 'glutes' that create the large bottoms of humans, and which help keep the body upright. The legs of hominins are longer than the arms, as a rule, and taper towards the ends. As anyone who has tried to walk while wearing big or mud-caked boots will know, smaller, lighter extremities allow for greater economy when walking than if the calves or feet were massive. The knees are held closer together, concentrating the mass near the vertical midline, so less energy is wasted in side-to-side movement. In addition, tendons in the legs recapture and reuse energy taken by each step to help power the next.[22] As a result, walking in humans is incredibly energy efficient. This explains why you have to walk a long, long way before you burn off any excess calories.

And this is just the start. Unseen, beneath the mechanical superstructure, is a system of nervous and muscular control of unrivalled precision, required to monitor the posture of the body at all times, relay these positional signals to the brain, and have the brain send out signals to correct the posture with split-second timing. This is because walking on two legs is far less stable than on four, as each step requires leaving only one foot on the ground, inherently unbalancing the entire body. The brain and body must get a good picture of where the body is in space and where it is going, simply just to stop it toppling over while it does something as seemingly simple as taking a step.

The health consequences of bipedality, however, are huge. Gravity imposes costs on a vertical structure balanced on just two legs that quadrupeds can largely avoid. Injuries to backs, knees, hips and feet exact an enormous toll: Chimpanzees and gorillas suffer much less from degenerative joint disease than humans.[23] Female humans tread a fine tightrope between a narrow pelvis required for efficient walking and a wide pelvis required for risk-free birth of a large foetus. The mass and shape of a female change constantly during pregnancy, leading to specialized adaptations in the lower back.[24] Pregnant females are engaged in a balancing act whose weights are constantly changing. The balance isn't perfect, as the foetus is born relatively early, compared with the babies of our closest ape relatives, and this can have consequences for health outcomes in later life, such as middle-ear problems.[25]

In all bipedal humans, of both sexes, and whether pregnant or not, bipedalism exacts a price, even after seven million years of evolutionary striving to make it as good as it can be. In a biped such as a human, whose mass is concentrated near the vertical midline, all the forces of walking – such as those transmitted through the contact of the feet with the ground – are passed directly up the legs and spine. This means that bipedality has a direct impact on many parts of the body, from the feet, up through the knees, hips and back, to the neck and head. Back pain remains one of the most frequently cited causes of worker absenteeism in the world. Even in physically fit bipeds whose backs don't hurt, the heart and blood vessels must work against gravity to ensure that blood flows evenly to all parts of the body and doesn't pool in the abdomen or feet. Failures in this system mean that bipedality might even be a root cause of some cases of hypertension[26] and varicose veins. From hernias to haemorrhoids, bipedality has a lot to answer for.[27]

Perhaps inevitably, then, the rapid evolution of such a thoroughgoing transformation as habitual bipedalism could never be anything but a somewhat Heath Robinson[28] arrangement. It is a testament to the power of evolution that humans are as good at it as they are, but the costs of bipedalism have scarred human evolution. Even before *Homo sapiens* evolved, the bizarre habit of bipedalism laid a burden on human life that it could never hope to shake entirely.

## 2

# THE GENUS *HOMO*

*Artaxerxes had served with great reputation in the armies of Artaban, the last king of the Parthians, and it appears he was driven into exile and rebellion by royal ingratitude, the customary reward for superior merit.*

EDWARD GIBBON,
*The Decline and Fall of the Roman Empire*

One thing I hope to show in this book is that humans – and hominins in general – like to rise to a challenge. Despite bipedality (and some might say, because of it), hominins evolved and diversified during the Pliocene epoch (roughly 5.3 to 2.6 million years ago), when the climate of Africa became drier, and woodlands were giving way to a more mixed environment of dry grassland and scattered stands of trees close to waterholes and rivers. During this period, hominins spread through Africa. Their remains are known from Chad in west central Africa, to the Great Rift Valley in the east, and down to the 'Cradle of Humankind' in South Africa.[1] They began to use tools. At first these were no more than chipped cobbles,[2] used for smashing bones to extract the nutritious marrow, or tenderizing meat and pounding vegetation to make it more palatable.[3]

Animal bones marked with the distinctive scratches that are the signs of deliberate butchery have been found that are more than three million years old.[4] The genus *Australopithecus* – a group of species of bipedal, opportunistic scavengers – was joined by another genus of hominin, *Paranthropus*.

Like the gorilla today, *Paranthropus* was a specialist vegetarian. It had massive teeth for grinding up roots and nuts, and big bellies to digest them in. *Paranthropus*, like *Australopithecus*, used rocks to tenderize raw food into more succulent fare. Unlike the gorilla, it was a native of the open plains and the occasional shady woodland of the drying Earth. But unlike other open-plains vegetarians – one thinks of zebra and gazelle, gnu, even baboons – *Paranthropus* was always rare, living in small, gorilla-style family groups. To be a vegetarian, and to live in small, marginal groups on the hardscrabble plains, was to draw all the low-value cards in the pack of life. Life became more and more tenuous for *Paranthropus* as the climate continued to dry out. *Paranthropus* disappeared from the fossil record just over half a million years ago. It's amazing they lasted as long as they did.

With Africa already drying out and the rainforests shrinking, the climate took another distinct turn for the worse around 2.5 million years ago. That was when a new hominin appeared in Africa – the first members of our own genus, *Homo*.[5] This was to be an altogether different proposition from *Paranthropus*. Rather than being a specialist vegetarian, *Homo* concentrated on eating meat. Not just scavenging, but actively hunting for it. In an increasingly hostile world, meat provided more calories in any given morsel than vegetable matter. But hunting required more smarts, too. Plants might be tough and thorny, but they don't run away. The first essays in the genus, though – creatures such as *Homo habilis* – were only mild variants of *Australopithecus*. Some of them might have taken one step backwards and two steps upwards and retreated into the trees.[6]

The new course towards a ground-living, pack-hunting carnivore was set by a creature known today as *Homo erectus*. A skeleton of a *Homo erectus* youth,[7] living in what is now Kenya 1.6 million years ago, reveals a dramatic contrast with *Australopithecus*. It was tall and long-legged, lanky and lean, quite unlike the somewhat squat and pot-bellied, long-armed and short-legged *Australopithecus*, or indeed earlier species of *Homo*. Its feet were firmly set on the ground. This time, there was no way back into the trees. Once completely grounded, it discovered a new gait. For, more than just walking, *Homo erectus* discovered that it could run.[8] And run it did, mile after mile, an increasingly erect and long-limbed gait accompanied by progressive hairlessness that allowed endurance running in tropical heat, to chase down prey that might have been faster over a short distance than its spindly pursuers, but which expired from heat stress as the miles accumulated.

*Homo erectus* was the first hominin that was more than a bipedal chimp or bargain-basement gorilla. It made beautiful stone tools. It tamed fire. It scratched signs on shells.[9] It made barbecues, maybe even boats. At the same time, it was not quite human. Its tools, though beautiful, show a kind of sameness and repetition that speak of instinctive rather than deliberate manufacture. And through all this it kept on running, running, following the game over the endless savannah that clothed almost the whole of Africa and much of southern Europe and western and southern Asia. *Homo erectus* was, as far as we know, the first hominin to have left Africa. Stone tools from China dated to just over two million years ago show that they ranged both far and fast.[10]

And, as it ran, it evolved. The earliest known skeletal remains of hominins outside Africa are 1.8 million years old and come from the Caucasus mountains,[11] and these hominins seem small, primitive. They soon bulked up, however. In Europe, *Homo heidelbergensis* stood as tall as modern *Homo sapiens* and might have been considerably beefier. As the climate deteriorated further and glaciers threatened to clothe the northern continents, hominins became more rugged, ranging as far north as eastern England.

The village of Happisburgh (pronounced 'Haze-boro') on the north Norfolk coast, in the east of England, has been attracting tourists for a long time. Sir Arthur Conan Doyle stayed at the Hill House Inn in Happisburgh, where he wrote 'The Adventure of the Dancing Men', one of his Sherlock Holmes stories.[12] When visiting north Norfolk, it's always good to have an indoor activity to hand, as the weather can be changeable, windy and cold. But humans have favoured this quiet spot for aeons. Some 800,000 years ago, some of the heirs of *Homo erectus* set up camp here.[13] It might have been a species called *Homo antecessor*, also known to be from Spain, and which lived around the same time.¹⁴ The rapid jags of climate change in the Ice Age meant that the prevailing cold of Europe was punctuated by short intervals of almost tropical weather. There were times when lions hunted game in what would one day become Trafalgar Square, in central London; hyaenas made dens in Yorkshire; and hippos wallowed in the River Tees. Significantly, though, the Happisburgh hominins were not fair-weather visitors. When they visited the area, the climate was no warmer than it is today: and if you want to exfoliate your face today, just stand on Happisburgh beach in January, facing into the wind.

Around 350,000 years ago, a group of *Homo erectus* could be found living in caves in northern Spain.¹⁵ They were much more heavily built than their African ancestors. They were on the cusp of becoming the first Neanderthals – the acme of the Ice Age caveman.

Ranging across Eurasia, the Neanderthals were heirs of a lineage that had until then preferred to live in the tropics. Ruggedly adapted to the cold, their adaptations ran to more than just a heavy, solid build. They started to modify their environments to survive. Rather than weather the glacial chill outdoors, they began to cultivate a deep inner life, in caves and rock shelters, emerging only to hunt. In their caves, they found a rich life of the mind – appropriate to the large size of their brains, on average larger than ours. They built structures inside their caves, far from daylight.¹⁶ They revered their elders. They buried their dead. These troglodytes were as far from the free-running, antelope-chasing *Homo erectus* as could be imagined. But although they survived for more than 300,000 years, they suffered the blight of all hominins – that is, they were never common.

## THE GENUS *HOMO*

With rather small home ranges, Neanderthal clans rarely interacted with one another. They became somewhat inbred.[17] They were forever just on the verge of buying a one-way ticket to oblivion.

Small populations spread thin are meat and drink for evolutionary change. As *Homo erectus* roamed across the continents, it evolved into a wonderland of different forms, and not just *Homo heidelbergensis* and Neanderthals. A million years ago, Eurasia came close to Tolkien's Middle-earth, in that it became home to a wide range of human-like beings. All were descendants of *Homo erectus*.

For there were giants on the Earth in those days. *Homo longi* – 'Dragon Man' – was discovered from a skull found in Manchuria in the 1930s. It came from a person at least as tall and massive as anyone living today.[18] It was one of several different human species living in eastern Asia between the arrival of *Homo erectus* more than two million years ago and *Homo sapiens* almost two million years later.

One species was the Denisovans. Named from Denisova Cave in the Altai Mountains of Siberia, where evidence of their existence first came to light, these humans, close cousins of Neanderthals, evolved in one of the most hostile environments on Earth outside the polar regions – the high plateau of Tibet. If any creature comes close to the mythical yeti or 'Abominable Snowman', it is this.

More curious still are the hobbits of island south-east Asia – *Homo luzonensis* from the Philippines[19] and *Homo floresiensis* from Flores, in present-day Indonesia.[20] These tiny people, no more than a yard or so in height, descended from bands of *Homo erectus* that made their way to these far-flung regions and became marooned.

For the shape of species is malleable and island life does strange things to them. Small creatures become larger, and large species become smaller. *Homo luzonensis* hunted small rhinos. *Homo floresiensis* lived alongside dwarf elephants the size of ponies and giant rats the size of terriers. But they stayed out of the way of Komodo dragons of a size much greater than even the formidable lizards of today: These hobbits shared their island with real-life dragons. But they made stone tools just like those of *Homo erectus*, only smaller.

The ancestors of *Homo floresiensis* came to Flores more than a million years ago. How they got there is a mystery. During the coldest parts of the ice ages, the world's glaciers and the great polar ice caps sucked up so much of the world's water that the level of the oceans was more than three hundred feet lower than it is today. This meant that what to us are islands were once connected to one another by low-lying land bridges. This explains how *Homo antecessor* could walk to Britain, and *Homo erectus* could walk to Java, without getting their feet wet. But Flores is separated from its neighbours by deep channels and has always been cut off. Either the ancestors of *Homo erectus* were blown there by chance (for such things do occur in nature) or they made boats. Once marooned, they lost their talent for boat building and evolution took its remorseless course. The last *Homo floresiensis* lived around 50,000 years ago, coincident with the arrival of *Homo sapiens* on the island, surviving – perhaps – only in local legends of Ebu Gogo, a secretive Little People of the remote countryside.

The most diverse continent in terms of ethnic tradition, language and genetics today is Africa – ultimately the home of all humans. It would not be true to say that when *Homo erectus* left Africa, it left its homeland deserted. Evolution continued there, too, with distinctive forms of hominin, though these are even less well known than the extremely varied hobbits, Dragon Men, yetis and troglodytes of Europe and Asia, of which we have tiny scraps of knowledge. *Homo rhodesiensis* is an ancient hominin that was discovered from a skull found in mine workings in what is now Zambia, and which lived there some 300,000 years ago.[21] *Homo naledi*, whose bones were discovered deep in a cave in South Africa, is of roughly the same age.[22] The first, though, was large and robust; the second, much smaller, and with a smaller brain and an anatomy more in tune with a hominin living much earlier. Then there are the skulls from Iho Eleru in Nigeria, which look archaic for all that they might be much more recent – less than 20,000 years old. It's clear that Africa played host to many different human species even as our own – *Homo sapiens* – was emerging there, more than 300,000 years ago.[23]

I cannot stress enough that even this varied menagerie of human forms that has come to our knowledge represents a tiny fraction of what might have been. For humans were always rare, and the processes of posterity fickle. There are some human species, now long gone, known more from the signatures of their DNA in modern humans than from actual physical remains. This is certainly true for the Denisovans: only a few scraps of teeth, a jawbone and tiny shards of bone survive from this entire species, too few to build up a reliable picture of what they looked like in life. But technology is such that the entire DNA sequence of a

Denisovan was read from one nondescript finger bone[24] – enough to show that Denisovans, through later interbreeding, left their mark in the genes of many people living today in Southeast Asia.[25] They bequeathed a gene that allows modern Tibetans to breathe easily in the thin air on the Roof of the World,[26] a vestige of the mountain ancestry of this remarkable hominin. And there are hominins even less well known than that – some genes carried by modern Africans suggest earlier interbreeding with species of human unique to Africa that have left no physical vestige of their existence at all.

The picture that emerges is of a group of varied species spread across vast areas, but thinly – so thinly that they would occasionally wink out altogether. For creatures living in very small groups, evolution exacts a pitiless toll.

There is safety in numbers. Creatures living in large groups can weather the occasional disaster. Small groups, however, become prey to random chance events. And there is a hidden, genetic cost to living in small groups. Inbreeding becomes a problem, and this can bring genetic disorders to light that in larger populations would have been extremely rare or stayed hidden.

Perhaps the most celebrated case of inbreeding occurred in the Habsburg dynasty,[27] which supplied Europe with kings and rulers for several centuries. Marriage between close kin was a way of keeping the various possessions within the family. Between 1450 and 1750, the Habsburgs celebrated seventy-three marriages, many of which were between near relations. For example, there were four marriages between uncles and nieces, two marriages between brides and grooms who were first cousins twice over, and one between partners who were second cousins four times over. When all Habsburg marriages are considered together, their degree of inbreeding was on average closer than a first-cousin marriage. Inbreeding leads to a general phenomenon called 'inbreeding depression', which, baldly, means an increase in the number of people who die in infancy or childhood or, in any event, before they can reproduce, compared with the population at large.

The most notoriously inbred clan of the sprawling Habsburg dynasty was the Spanish branch, which occupied the Spanish throne between 1516 and 1700. The last Habsburg king of Spain was Charles II, known as *El Hechizado* ('The Bewitched') by virtue of the large numbers of diseases and disabilities he suffered. Stunted, weak and thin, but with an abnormally large head, he could not speak until he was four, or walk until the age of eight. He suffered from repeated bouts of diarrhoea and vomiting, and as he got older, he was additionally crippled by episodes of hallucinations and convulsions. He died at the age of thirty-eight but looked older. Like many of the Habsburgs, he had an unusually protruding chin and lower jaw, which meant that his teeth did not meet properly, and he had trouble eating. He was married twice. Neither union produced children. On his death the Spanish Habsburg dynasty became extinct, leaving Spain without a king. In case one thinks that inbreeding doesn't have consequences, this event prompted the War of the Spanish Succession (1700–1714), a global superpower conflict that drew in sixteen countries and led to 400,000 deaths from combat and more than a million from disease.

It's impossible to know how much of poor Charles's problems were caused by inbreeding – we'd only know that if we had access to his DNA – but the tangled web of his family tree is telling. His father, Philip IV of Spain, was also the uncle of his mother, Mariana of Austria. Mariana was in turn the daughter of Holy Roman Emperor Ferdinand III and his wife Maria Anna of Spain, who was also Ferdinand's first cousin as well as the sister of Philip IV, Charles's father. Philip IV was, in turn, the son of Philip III and his second cousin Margaret of Austria, who was also the sister of Emperor Ferdinand II. This Ferdinand married his first cousin Maria Anna of Bavaria, the child of a union of second cousins, and their offspring was Ferdinand III. Ferdinand II was in turn the offspring of an uncle–niece union, as was Philip III. Charles's pedigree looks less like a genealogy than a ball of wool attacked by a litter of kittens.

It is possible for a highly inbred population to weather the toll of genetic diseases, because, eventually, all the people carrying genetic disorders will have died out without leaving any descendants. This appears to have been the case with the Habsburgs, who seem to have been prone to inbreeding depression more in the earlier part of their centuries-long dominance (1450–1600) than the later period (1600–1800). If the disorders are recessive, however, this gradual diminishment can take a very long time. The emotional and personal cost is barely imaginable. Even then, if the population has managed to avoid extinction, the resulting group, though free of disorders, will become remarkably similar. Finding mates outside the group will be even more essential, else the entire, rather homogeneous clan gets wiped out by some newly emerged disease, or a similar catastrophe.

Or it dies out, simply because the surviving members are too sickly to provide heirs that survive to reproductive age, or, in Charles's case, at all.

It should be said that the case of the Habsburgs is only exceptional in that, being the pedigree of the pre-eminent royal family in Europe, it is well documented. Consanguineous marriage (reckoned as being between people who are second cousins or closer) is common in many parts of Africa and Asia. A study of 5,000 army families in Pakistan, for example, revealed that more than three quarters of marriages (77 per cent) were consanguineous, with no fewer than 62.5 per cent being between first cousins – despite the increased toll of congenital defects in the offspring, compared with children of parents who were not closely related.[28] This is not an isolated case. It is only in contemporary Western societies that consanguineous marriage is frowned upon, and presumably only since Western science and medicine have become more aware of the mechanisms of heredity. It is perhaps an irony that Charles Darwin, father of evolution – who was indeed aware of the effects of inbreeding – married his first cousin Emma Wedgwood. The Darwins and the Wedgwoods were interlinked by a whole series of marriages, some of which were consanguineous. It is possible that the high mortality of the couple's progeny (three out of ten died in childhood) was related to inbreeding depression – the possibility of which Darwin was, poignantly, only too well aware.[29]

For early hominins, finding mates outside the group would have been vital for survival. In all primates, females tend to mate with males outside the group, and go to live with their mates' clan. As far as is known, this was true of

*Australopithecus*[30] as well as Neanderthals.[31] Small groups would meet at times of festival to sing, dance, swap tall tales and, most of all, marry off their eligible youngsters. But if groups become spread too thinly, too far apart such that they lose all contact, they are thrown on their own resources and blink out, one by one, until they are all gone.

In Africa, at around the same time that Neanderthals were beginning to emerge in Europe, our own species, *Homo sapiens*, appeared on the scene. The earliest evidence for this comes from Morocco and is around 315,000 years old.[32] At the time, *Homo sapiens* was just one of many. Even in Africa, other species, such as *Homo rhodesiensis*, *Homo naledi* and maybe many more, were scattered like thinly salted grains across the face of that vast continent. Who could have told, at the time, that *Homo sapiens* would one day come to dominate the entire Earth, driving all other hominins – and much else – to extinction?

# 3

# LAST AMONG EQUALS

*If the ninth and tenth centuries were the times of darkness, the thirteenth and fourteenth were the age of absurdity and fable.*

EDWARD GIBBON,
*The Decline and Fall of the Roman Empire*

For millions of years, meagre, precious scraps of fossil bone are nearly all that the record can tell of human evolution. With the evolution of our own species, *Homo sapiens*, there is other evidence to call upon – the genetics of people living today.[1]

Our genes are the direct record of our heritage. Although their use to infer ancestry is fraught with problems, it is still possible to detect signs of history written in the genetic code of people alive today. This was, in essence, the story revealed by a study in 1987 by the late Allan Wilson – a pioneer of modern genetics – and his colleagues Mark Stoneking and Rebecca Cann. Their method relies on the known property of genetic material to change, by mutation, over the course of evolutionary time, at a rate that can be estimated. By cataloguing the similarities and differences between the genetic material of 147 living people, they showed that the ancestry of all modern humans could be traced to Africa, and, specifically, to a woman living around 200,000 years ago. It wasn't too much of a stretch to give her a name – 'Eve'.[2]

But before we get too carried away with powerful Biblical metaphors, let's backtrack slightly. First, what is the genetic code?

Deep in the heart, or nucleus, of each one of the trillions of cells[3] in your body is a kind of minuscule book, containing the instructions for the health and maintenance of cells, tissues and bodies. The book takes the form of a long, string-like substance called DNA, composed from an alphabet of four chemical letters called 'bases', arranged in sequence to make chemical sentences. Each sentence in this book is a gene. There are about 30,000 genes in your recipe book, written in more than three billion bases.[4] Each gene comes in two copies – one inherited from your father, the other from your mother. This is your 'nuclear' DNA.

But each cell contains more DNA – thousands of copies of a much smaller, 16,569-base segment called 'mitochondrial' DNA, contained within bodies in the cell called mitochondria, outside the nucleus. Mitochondrial DNA is inherited exclusively from your mother. Wilson and his colleagues reconstructed their human genealogy based on mitochondrial DNA, so had traced, therefore, not a family tree as such, but a matriline, a sequence of mothers. Hence 'Eve', and, more specifically, 'mitochondrial' Eve. They showed that all mitochondrial DNA sequences taken from people of European or Asian descent were offshoots of a much larger tree that was exclusively African. This was an indicator that the ancestress of all people outside Africa – and all modern people inside Africa as well – would have been African.

Eve, however, was not the only woman. There would have been many other women – mothers, sisters, daughters – but they would not have left descendants, or, at least, descendants that lived to the present day. Extinction has exacted such a toll on humanity throughout the ages that almost all lineages that once existed have died out. This would give the impression, after the fact, of a single ancestress of human beings. Neither was Eve the first. She was herself the descendant of uncounted generations of hominins distinguished only by being the last common ancestress of all modern humans.

Nor would Eve herself have been of any special importance at the time. She didn't have a halo, or a sign on her head to say that she, of all her sisters, would be the only one to leave descendants 200,000 years later. Perhaps she had one, tiny, selective advantage, too small to see at the time, that would pay out over the generations. Or maybe she just got lucky.

Here is a modern thought experiment to set this into perspective. Imagine that today there is a woman living in Africa who just happens to be naturally immune to type 1 human immunodeficiency virus (HIV-1), the virus that causes acquired immunodeficiency syndrome (AIDS). This small advantage could play out over the generations such that, in thousands of years' time, all humans could number her among their ancestors.[5] But she wouldn't know this. Neither would her best friend, also immune to HIV-1, who, for whatever random reason, dies childless and so leaves no descendants at all.

The study of Wilson and his colleagues was limited, based on the limited technology of the time, when it was possible to read only the relatively short recipes encoded in mitochondrial DNA. Since then, however, it has become possible to read nuclear DNA. On 12 February 2001, The Human Genome Project announced that the entire sequence of nuclear DNA in the human genome – all three billion bases – could now be read. In those days, the multimillion-dollar, multinational effort was compared with the Apollo Project to land living human beings on the Moon in the 1960s. More than two decades later, thanks to rapid advances in gene technology, reading nuclear DNA has become almost commonplace. The findings generally confirm the revelations of the Wilson group. That is, modern humans all descend from people who came from Africa.

But if there was an Eve, did she not have her Adam? A section of nuclear DNA called the Y chromosome is always passed down from father to son. Tracing the history of the Y chromosome shows that there was indeed a 'Y-chromosome Adam' – a patrilineal common ancestor of all people now living – but he lived at a much longer remove than Eve, possibly at the time that *Homo sapiens* first emerged as a species. Y-chromosome Adam and mitochondrial Eve would never have met.

The Wilson paper was published just before I joined the staff of the science journal *Nature*. I was still a graduate student at the University of Cambridge but spent a lot of time working in the collections of the Natural History Museum in London. There I met and got to know Chris Stringer, an expert on the origin of our own species. He had long championed the 'Out of Africa' view, that modern humans (that is, *Homo sapiens*) could all trace their ancestry to Africa. Humans emerged from that continent and replaced all the other hominins living in the world, so that only *Homo sapiens* remained. Ranged against this view was the 'multiregional continuity' view – humans evolved in many places across Eurasia, but eventually came together as a single species by a process of interbreeding. I remember travelling on the London Underground with Chris and discussing how much the Wilson work vindicated his own ideas, based on comparing the fossil remains of ancient hominins. The situation has since become more nuanced. Yes, modern humans did originate from Africa, but did not completely replace the hominins already living in other parts of the world. In some cases, they interbred with them.

Behind every person now living, wrote the late Arthur C. Clarke in *2001: A Space Odyssey*, stand thirty ghosts. In a way, studying the genes of people invokes the shades of our ancestors, but cannot clothe them in flesh. This is where the fossils come in – with the proviso, frustratingly, that fossils in and of themselves need not represent the ancestor of any modern human.

The earliest fossils that can be placed in our own species, if only just, are several skulls from a site in Morocco called Jebel Irhoud.[6] The skulls look heavy-set, and antique compared with those of modern humans. Modern skulls are tall and globular, with small faces tucked in under the forehead, a distinct bony chin and modest ridges beneath the eyebrows. The Jebel Irhoud skulls are almost like this, but not quite, and some look distinctly Neanderthal-like – long and low, with large faces and distinct brow ridges. The people to whom these skulls belonged lived approximately 315,000 years ago,[7] at about the same time that Neanderthals themselves were emerging, and when other species of hominin such as *Homo rhodesiensis* and *Homo naledi* lived in Africa. It is possible that both Neanderthals and humans can trace their ancestry to something or someone close to *Homo antecessor*, which lived around 800,000 years ago.[8] Its fossil remains show a primitive skull shape but a remarkably human-like face. If we were to pin a name to the species that left its footprints[9] and stone tools[10] at Happisburgh in Norfolk, it might well be *Homo antecessor*.

A hundred thousand years or more were to pass before *Homo sapiens* left any more corporeal calling cards. These were two skulls from Kibish in Ethiopia, one of which looks more human-like than the other, both about 200,000 years old;[11] and a further skull, perhaps 160,000 years old, from Herto, also in Ethiopia, which looks more like that of *Homo sapiens* than anything else, but which still has a more archaic cast.[12] From all this it seems clear that *Homo sapiens* started out as a somewhat raw ingredient, to be slowly refined into shape by evolution. A hundred more millennia were to pass before *Homo sapiens* appeared that would pass unnoticed in a crowd of modern people.

The lessons that emerge from the vanishingly minimal remains of humans in Africa during the earliest phase of the evolution of our own species are, first – and once again – that humans were always rare. It is not that fossils themselves are uncommon from this time. The fossil record of Africa at the time groans under the weight of remains of antelope, wild pigs of various kinds, fishes and crocodiles – much like the typical African fauna of today. When I visited a palaeontological team prospecting for fossils on the western shore of Lake Turkana in 1998, exploring sediments laid down more than three million years ago, the fossils of catfish, turtles and crocodiles were too common to be bothered with. Fossils of pigs and antelope-like animals were worthy of some passing notice. But the entire haul of hominin fossils – almost all fragments of teeth – could be housed in a (very small) tin box.[13]

Second, the early fossils of *Homo sapiens* show extraordinary variability, in both time and space. This speaks to a species living in small groups spread thinly across the landscape. Each group would have its own tics, its own quirks, its own peculiar family resemblances, its own susceptibilities to disease. But, as in all primates, males would seek females from different groups, and this occasional genetic interchange would stop the whole fragile social network fragmenting into disconnected webs and patches, each becoming more and more inbred until, one by one, they died out.

Third, and most importantly for the purposes of this book, animals living in small groups are prey to extinction by random events, in addition to the burden of inbreeding. A flash flood might wash them all away. And, in a drying climate, where game animals were increasingly hard to find, they were always skirting the edge of extinction through drought or famine.

These early, unformed examples of our species did occasionally venture out of Africa. All such adventures ended in extinction. One of the earliest excursions was to Europe – a skull from Greece suggests that *Homo sapiens* had a toehold there some 210,000 years ago.[14] The Levant, just outside Africa, saw the occasional early human presence flickering and guttering several times, only to retreat, before occupation became more permanent. Caves in the Mount Carmel massif in Israel show infrequent intrusions of modern humans from Africa into a Eurasia dominated by Neanderthals, as evidenced by human remains dating to more than 90,000 years ago in caves such as Skhul and Qafzeh.[15] Early dispersals round the Indian Ocean basin led to the appearance of *Homo sapiens* in Australia as early as 65,000 years ago,[16] as well as Sumatra (73,000–65,000 years ago)[17] and China (more than 80,000 years ago),[18] though these populations almost certainly died out, to be replaced by separate, later expansions from Africa.

Everywhere it went, *Homo sapiens* encountered its cousins, inheritors of the much earlier migrations of *Homo erectus*. There were other species living in Africa, perhaps until recent times, geologically speaking. Outside Africa they met Neanderthals in Europe and western Asia, and Denisovans in east and south-east Asia. And where they met, they interbred.

Genes from modern people can be used to trace ghostly family trees into the past. Fossils reveal what early humans looked like, but bones alone cannot say much that is specific about ancestry – any fossil human you might uncover *could* have been your direct ancestor, but it's impossible to know for certain.[19] The most that a fossil can be is your cousin to some degree. However, the technology of gene sequencing means that it is possible to extract DNA from fossil bones themselves, and in recent years this has opened a whole new window on human ancestry.

The extraction of a complete mitochondrial DNA sequence from a Neanderthal by Svante Pääbo and colleagues, just ten years after the mitochondrial 'Eve' paper by the Wilson group, was a milestone.[20] (Pääbo went on to win a Nobel Prize for his pioneering work on recovering so-called ancient DNA.) Since then, many whole nuclear sequences (or 'genomes') have been read, showing that Neanderthals lived in small groups and were rather prone to inbreeding.

The technology of handling small, ancient and fragile pieces of ancient DNA has advanced massively since the pioneering work of Pääbo and his colleagues. It is now possible to extract and decipher nuclear genomes from pieces of bone too small and nondescript to be assigned to any particular species based on their external shape. Thus, it was a sensation that DNA extracted from a shard of finger bone, no larger than a grain of rice, from a hominin that lived in a cave called Denisova in the Altai Mountains of Siberia turned out to have come from a hitherto unknown species.[21] The ancestry of this species, known informally as Denisovans, diverged from the Neanderthal lineage after their common lineage diverged from that of *Homo sapiens*. Denisovans and Neanderthals were first cousins.

The shocker is that modern *Homo sapiens* is a product of interbreeding with both Neanderthals and Denisovans, as well as other hominin species that remain undiscovered except for signs of their passing hidden in the DNA of people alive today.

After a quarter of a million years of evolution in Africa, with only occasional migrations that ultimately failed, *Homo sapiens* was ready to emerge. A small group of *Homo sapiens* – or perhaps several small, closely related clans – migrated from Africa sometime between 60,000 and 40,000 years ago, eventually to replace every other human and hominin group throughout the world.

*Homo sapiens* colonized Europe from the south (through Italy) and east (along the Danube, through modern Romania and Bulgaria)[22] around 47,000 years ago. On the way they met Neanderthals and occasionally interbred with them. The DNA from an early modern *Homo sapiens* that lived in Romania about 40,000 years ago shows that it had a Neanderthal great-grandparent.[23] This does not seem to have been an isolated incident.[24] Today, about 2 per cent of the DNA in every human who does not have an exclusively African ancestry comes from Neanderthals. It would once have been more but for natural selection, which has since weeded out any genes from Neanderthals that were likely to be deleterious or mutated. Because Neanderthal populations were always small, the load of bad mutations was rather high.

Denisovans contributed a small smattering of DNA to modern humans, in particular those in east and south-east Asia, such as the gene that allows modern Tibetans to survive and even thrive at high altitude.[25] As noted before, there are intriguing signs in modern human DNA of interbreeding, long ago, with hominins that are now extinct and have no known fossil record, whose existence is known only from the DNA they have contributed to modern humans.[26] Some Denisovan genomes suggest signs of interbreeding, perhaps a million years ago, with a still more ancient hominin species. This might have been *Homo antecessor*.

Only a small fraction of the humans that originally evolved in Africa migrated away from it. The work on mitochondrial DNA by Allan Wilson and his colleagues showed that all modern humans outside Africa represent just one offshoot of a much more diverse picture within Africa itself. Africa is, and always was, the most diverse continent as regards the shape and form of human evolution. *Homo sapiens* seems to have emerged from a gradual pulling together of a range of rather diverse forms over a long period and appears to have been a species-in-waiting for a very long time. All the while, climate change exacted a toll on small, widely scattered groups of varied peoples. Long periods of arid conditions (corresponding with ice ages in Europe) would have broken connections between groups of people, allowing each to evolve in its own way, whereas brief, warmer interludes might have allowed moments of reunion and respite.[27]

The relationships of modern humans with more ancient varieties within Africa are wreathed in shadow. For example, does the ancient hominin represented by *Homo rhodesiensis*, dated at around 300,000 years ago,[28] have anything to say about human origins? The fossils are mute, so we are constrained to tell their stories on their behalf. But what stories should we tell? The clues we have are, as usual, few, and enigmatic, but speak of a diversity of human form in Africa that was once greater than it is today.

Around 20,000 years ago, when the most recent ice age in Europe and North America was just past its peak, a community of fisherfolk lived at a place now called Ishango in the present Democratic Republic of Congo.[29] Although this is but an eyeblink ago in terms of geology, the people from Ishango were unusual by modern standards, with thick, heavily built skulls more reminiscent of much more ancient people from the Levant, as exemplified by the early *Homo sapiens* remains from Skhul and Qafzeh. At around the same time, there lived a group of people around a place now called Lukenya Hill in Kenya, whose skulls recalled, in their shape, thickness and the presence of big brow ridges, a much earlier period of prehistory.[30] Slightly earlier – between 40,000 and 35,000 years ago – the skeleton of a young miner was buried at a chert mine in what is now Egypt. Robust and heavily built for a life of hard labour, the young man also had the heavily browed skull of an earlier age.[31] Even more recently, perhaps 14,000 years ago, a person was buried at Iho Eleru in south-west Nigeria, about 125 miles east of the place where the megacity of Lagos would one day be built.[32] This person had a long, low skull, more like that of the early *Homo sapiens* from Skhul and Qafzeh – or the fisherfolk of Ishango – than someone living on the cusp of, what are in geological terms, modern times. It seems that, all over Africa, people were popping up whose shapes were shifted from what we would understand as modern, showing either that the range of human forms in Africa was once greater than it is today, or that ancient forms of human lived in Africa until relatively recently.

It seems that the DNA of modern Africans contains a small contribution – about 2 per cent – from an otherwise unknown hominin species that became distinct from the ancestors of modern humans about 700,000 years ago, rather in the same way that the DNA of modern non-Africans contains some from Neanderthals and Denisovans. This DNA made its way into the ancestors of modern Africans around 35,000 years ago,[33] which makes sense given the archaic form and recent date of human remains from Iho Eleru, Ishango and elsewhere.

The most divergent human lineage in the world today belongs to the KhoeSan people of southern Africa. Study of ancient DNA from Africa suggests that the hunter-gatherer KhoeSan were much more widespread in the past, having largely been replaced in their former range by pastoralist herders. However, there is a trace of an ancient ancestry in west Africa,[34] all but erased by subsequent population movements. The story is that Africa retains a deep and complex structure only lightly sampled by the tiny population that left between 60,000 and 40,000 years ago to make its way into the rest of the world, expunging almost all traces of the once-diverse range of hominins that lived there. Yet the diversity of Africa is nowhere near what it was as recently as 20,000 years ago. The homogenization of *Homo sapiens* happened even there. Apart from a few fast-disappearing outliers, by around 50,000 years ago, *Homo sapiens* was the last of the hominins, the Last Human Standing.

# 4

# LAST HUMAN STANDING

*Commodus killed a camelopardalis or giraffe, the tallest, the most gentle, and the most useless of the large quadrupeds. This singular animal, a native only of the interior parts of Africa, has not been seen in Europe since the revival of letters, and though M. de Buffon (Histoire naturelle) has endeavoured to describe, he has not ventured to delineate, the giraffe.*

Edward Gibbon,
*The Decline and Fall of the Roman Empire*

For almost the entirety of its history, *Homo sapiens* has been extremely rare. Tiny bands of people, spread thinly over the Earth's surface, living a subsistence existence barely a meal from starvation, two meals from extinction.

There is another disadvantage in scarcity – other than closeness to oblivion – and that's the general absence of brain power. It takes a village to raise a child, it is said, but it takes a civilization of millions to create a literate, technological civilization. Isaac Newton devised his theory of gravity when he was one of 500 million people. In 1905, the year Einstein published his special theory of relativity, the world population was more than 1.6 billion. For 96 per cent of its existence, *Homo sapiens* has scratched out a living of sorts with basic Stone Age technology. Such technology only permits a small population to exist at any one time – a population that could disappear at any moment. Larger populations open the possibility of continuity, such that the inventions and traditions of one generation might be passed on to another, and another still, without these traditions vanishing along with the people, and having to be reinvented from nothing. Larger populations also mean more brains, more conversations, more exchange of ideas, more inventions. When Newton said that he could only see further because he stood on the shoulders of giants, his metaphor ran more deeply than even he knew.

Only after the invention of agriculture, and a more settled life, could populations accumulate in such a way as to permit such exchange of ideas. Suddenly, things appeared that had been rare or completely absent in the long span of prehistory – metalwork, pottery, the written word. To be sure, settled life had its costs (epidemic disease, large-scale war), but without it, and the expansion of population that it fostered, *Homo sapiens* would have been a hunter-gatherer forever, living in small populations where the accretion of new ideas would have been painfully slow.

In a cave on the southernmost shore of Africa, a human being used a crayon made of red ochre to scrawl a cross-hatched pattern on a fragment of rock. This is the first known example of drawing in history, and it is 73,000 years old.[1] It pre-dates the next-earliest examples of drawing – from cave paintings in Europe and Indonesia – by 30,000 years. The rock fragment comes from a place called Blombos Cave and is not an isolated case. It is part of what archaeologists call the 'Still Bay technocomplex', a flowering of prehistoric culture in southernmost South Africa between about 77,000 and 73,000 years ago. Not too far away, at a site called Pinnacle Point, is evidence that people were making tiny arrowheads from heat-treated stone at about the same time.[2] The use of pigment, personal adornments (such as beads made from shells), sophisticated tools made from bone, and projectile weapons such as bows and arrows are all developments that archaeologists regard as signs of modernity.

But these early signs of such behaviour were not the first.

There are signs of modernity at Blombos Cave from layers dated to 30,000 years *earlier* – as long as 100,000 years ago. These include pieces of ochre with cross-hatched engravings, as well as abalone shells that were used as artists' palettes to mix ochre pigment.[3] But the layers at Blombos Cave between the pigment factory and the Still Bay technocomplex show no signs of engraved ochre. It's as if the technology disappeared for 30,000 years – six times as long as the entire span of recorded history – and had to be reinvented.

It could be that the tradition of modernity was carried on in places other than Blombos Cave, but they have either not been discovered yet, or left no traces for us to find. But one must ask the question of why early signs of modernity at Blombos seemed to have faded out at all, rather than carrying on. It could be that the people there died out or moved on in the intervening millennia. Because of the immense spans of time involved, local extinction seems more likely, considering that the odds that small, isolated populations will persist for long periods of time are minuscule.

In addition, given the span of time, one is entitled to ask why – even if the culture did persist, unseen, between 100,000 and 70,000 years ago – it did not advance materially. For an interval of 30,000 years, nothing happened. To compare these ancient cultures with the modern age is instructive. Here, for example, is a personal anecdote. My Uncle Henry[4] was born in Potsdam, in the German Empire, in 1904, a year after the Wright brothers first flew their powered box kite on the Carolina coast. He was only sixty-five at the time of the first manned landing on the Moon (Apollo 11, 1969). But the expanse of time between the ochre factory at Blombos and the Still Bay technocomplex was almost five hundred times as long, and yet the advancement in technology between them is scarcely measurable. There is no sense that Stone Age people were unintelligent. Their brain capacities were every bit as large as those of people today. So, what was holding them back?

Again, the answer is the small population size. Populations were just too small, too thinly spread and too likely to disappear for technology to persist. New ideas, new inventions were lost before they could become established, and had to be reinvented repeatedly before they could eventually take root. The passage of some technical skill from teacher to learner, over generations, requires there to be a minimum number of people, living sufficiently close together, for that skill to become established – and that number and population density go up markedly with the degree of difficulty of that technical skill.[5] Mixing ochre to make pigment might require a few dozen skilled people and willing apprentices. Rocket science, on the other hand, requires a civilization of billions. Nobody knows when the crucial moment of technological take-off happened, but it was the moment when populations became large enough, or sufficiently well connected, for traditions to be maintained, and perhaps developed. Even then, progress, by today's standards, was infinitesimally slow.

It was enough, though, for *Homo sapiens* to – finally – break out of Africa.

It's important to realize that our ancestors had no concept of a continent called Africa, nor that they were leaving it. All they were doing was following where the wildlife they hunted led, or the plants they harvested grew, and setting up home in places that suited them. *Homo sapiens* started out as a species living in tropical savannah grassland and mixed woodland near water. It tended to avoid places that were either too dry (hot deserts) or too wet (tropical rainforest). As Robin Dennell says in *From Arabia to the Pacific: How Our Species Colonised Asia*, it's an irony that the people most affected by the ice ages of the past million years or so never saw a lump of ice nor felt a snowflake. *Homo sapiens* evolved in the open grasslands that clothed much of Africa and Eurasia during most of the ice ages, except for the times when ice was at its maximum in the far north, when the savannah dried out into desert; or the briefer interludes during which the ice melted and sea levels rose, when even the Sahara and Arabia were green and well-watered.[6] In terms of climate and habitat, Africa was continuous with Arabia, the Levant and south-west Asia as far east as the Thar Desert of north-west India. The first people to migrate to these regions, says Dennell, didn't leave Africa so much as take Africa with them.

Although *Homo sapiens*, in its broadest sense, originated in Africa and spread all over the continent rather quickly (as well as its Arabian and Asian extensions), it's perhaps a surprise to learn that it didn't spread sooner than it did. At issue was a kind of climate switchback. In the coldest parts of the ice ages, sea levels sank by as much as 130 yards, allowing humans to walk to otherwise inaccessible islands across land bridges previously submerged by water, or cross what were once deep estuaries simply by walking over the dry land so exposed. They could always walk into the Levant via the Sinai Peninsula, but another route was across the narrow strait of Bab el Mandeb between Africa and Arabia at the southern end of the Red Sea. At its narrowest it would have been less than three miles wide, the far shore easily visible to a tall person on the hither side. But at these favourable times, the climate of Africa was hot and dry, and deserts such as the Sahara would have been a deterrent to migration. At the opposite extreme, when the Sahara and Arabia were lush and green, sea levels would have been higher than they are today, flooding what would once have been easy migration routes. In the event, humans could only migrate during those sweet spots when the climate was not so dry that it became desert, nor so wet that migration routes were barred by water.

There was another barrier to migration, at least at first – the destinations were already occupied. As a result, the first essays in global conquest came to naught. As I noted above, small bands of *Homo sapiens* made it to Greece some 210,000 years ago, and Israel 90,000 years ago – at times when the climate and habitat were extensions of the African homeland – but did not become established, and almost certainly disappeared like frost in summer. In those days, the Neanderthals had Europe locked up tight. Climate, and the lie of the land, favoured migration from Africa along the Indian Ocean littoral, and into south-west Asia, where the Neanderthal realm petered out.[7] Judging by the fact that all modern humans whose ancestries are not exclusively African have a small amount of Neanderthal DNA, not all contact between *Homo sapiens* and Neanderthal was hostile. However, rather in the same way that the earliest English colonies of Virginia and the Carolinas failed for lack of people, disease and problems of resupply, these first essays in the migration of *Homo sapiens* from Africa contributed little or nothing to the modern human gene pool.

The Neanderthal barrier in Europe might explain why *Homo sapiens* broke out of Africa into southern Asia before they entered Europe. Evidence, however, is sparse. There are suggestions that *Homo sapiens* arrived in India more than 300,000 years ago – to judge from a set of stone tools excavated from a site called Attirampakkam in the far south of the peninsula,[8] though these artefacts could have been made by indigenous hominins such as *Homo erectus* that had made their way out of Africa much earlier. The general view is that *Homo sapiens* migrated round the Indian Ocean basin somewhat later.

It's important to note that not all these migrations

originated in a single wave, as if they were the Biblical Israelites crossing the Red Sea. Migration out of Africa and across southern Asia doubtless happened many times, and there would have been significant reverses. After the spectacular eruption of Mount Toba in Sumatra some 70,000 years ago – the largest volcanic eruption of the past two million years, and coincident with climatic deterioration at the end of the most recent interglacial – humans in southern Asia might have been cut off from those in Africa for tens of thousands of years.

*Homo sapiens* differed from earlier hominins that left Africa – notably *Homo erectus* – in its willingness to deviate from the ancestral hominin preference for living in tropical, open, savannah grasslands. To be sure, Neanderthals had long become used to the severe cold of northern Europe, and hominins had lived in England some 800,000 years ago, when the climate was no less cold than it is now. A particularly challenging environment, however, was tropical rainforest, which hominins had taken pains to avoid – until *Homo sapiens*. Although rainforest is rich in biological diversity, it's hard to make a living there. Rainforest is about the worst place to live for a species used to chasing large game over open country. Most animals in rainforest are small, difficult to see (all that vegetation gets in the way), live in hard-to-reach places such as the treetops, and may be venomous. Although rainforest has a cornucopia of plants, many of its staples are poisonous and cannot be eaten unless detoxified, sometimes using long and complex recipes. And yet *Homo sapiens* was living in rainforest in Sri Lanka and island south-east Asia over 40,000 years ago – something that no hominin species had done before. As Robin Dennell has pointed out, *Homo sapiens* is a uniquely invasive species,[9] boldly going (as somebody once said in another context) where no one had gone before.

Such venturesomeness was augmented by the invention of watercraft, which would have been essential to getting across the eastern parts of what is now Indonesia and to Australia. For, even when sea levels were at their lowest, many islands were not joined to the mainland. During such low-water episodes, Borneo, Sumatra and Java were joined to the mainland of south-east Asia, and a corridor of savannah grassland ran all the way down from southern China through Malaysia and into Java. At the same time, Australia, New Guinea and Tasmania were joined together in a large land mass. However, the many islands between – such as Sulawesi and Flores and the Moluccas – remained as islands. When sea levels rose, however, the land masses shrunk dramatically, forcing *Homo sapiens* to make much more use of marine resources than it had been used to – as attested by early evidence of fishhooks, and the bones of large marine fishes in archaeological sediments.

Sulawesi was always an island, but around 40,000 years ago it was witness to a remarkable flowering of culture that included cave art – paintings of large animals daubed in ochre on cave walls.[10] Similar cave art has been found in Borneo[11] and appeared at about the same time. This artwork is about the same age as, or older than, the more famous cave paintings of Lascaux in France and Altamira in Spain. From Blombos to Borneo, Still Bay to Sulawesi, wherever *Homo sapiens* went, it made its mark.

So, by 40,000 years ago, or thereabouts, *Homo sapiens* had penetrated island south-east Asia, and had even made landfall in Australia. It had learned to live in hostile rainforest, and had begun to make appreciably long journeys over sea by boat.

At the same time, other bands of *Homo sapiens* had migrated into the interior of central Asia. Although the steppes of southern Russia, Mongolia, northern China and Siberia had more in common with the ancestral African homeland than rainforest – wide open spaces with plenty of large game animals to chase – they offered their own challenges, principally extreme cold. *Homo sapiens* could not have survived the harsh environment of central Asia without shelter, fire and warm clothing.

The invention of clothing has left its mark in evolution, though not so much in humans as in their parasites. The human head louse (*Pediculus humanus capitis*) lives exclusively on the scalp; the body louse (*Pediculus humanus corporis*) infests regions lower down and makes its home in clothes. The two species share a common ancestry and went their separate evolutionary ways roughly 70,000 years ago – coincident with the time *Homo sapiens* left Africa and a reasonable fit for the time that clothes might have been invented.[12] Because of clothes, humans managed to penetrate as far north as the Arctic Ocean. The draw was the woolly mammoth – a source of meat, fat, hides and bones for tools, even houses. The prize must have been worth the (tallow) candle given the extreme cold and winter darkness: Neanderthals, for all their ruggedness, never penetrated further north than 60 degrees, well south of the Arctic.

From Arabia to Japan, from Australia to the Arctic, *Homo*

*sapiens* drove any indigenous hominins it met to extinction (*Homo floresiensis* in Flores, Indonesia; *Homo luzonensis* in the Philippines; perhaps the last vestiges of *Homo erectus*) or assimilated them into the human gene pool as they went (Neanderthals and Denisovans). It took a little while longer for *Homo sapiens* to penetrate the Neanderthal heartland of Europe, but after many false starts, this, too, was achieved. *Homo sapiens* left artefacts no Neanderthal could ever have produced – sophisticated bone tools, needles for sewing, exquisite cave paintings, and portable art including the repeated sculptures of obese, huge-breasted women, made from mammoth ivory – invocations to a higher power to promote fertility and stave off starvation. *Homo sapiens* occupied places that no other hominin ever had – notably Australia, the Americas (from at least 20,000 years ago), and eventually the more distant oceanic islands.

Why was *Homo sapiens* so successful an invader, leaving all the other hominin species in the shade? Nobody knows, but the advantages were likely to have been incremental. After all, for 86 per cent of its existence – from 315,000 to about 45,000 years ago – *Homo sapiens* was just another human species on the landscape. It's not that some magical switch clicked from OFF to ON 45,000 years ago. Rather it was a slow, exponential accumulation of changes in biology, behaviour and population starting maybe 100,000 years ago that resulted, tens of thousands of years later, in a situation in which *Homo sapiens* had occupied the entire world.

It is the nature of exponential growth that it starts small and is small for a very long time, until the slope steepens with dramatic suddenness. The change from slow crawl to sudden climb happened around 45,000 years ago, when human behaviour was essentially modern; human populations had expanded to a degree that allowed technology to become established and to develop; and for all other hominins to have been competed out of existence.

It is perhaps worth exploring why the other hominins, particularly Neanderthals, couldn't stand the competition. After all, they had been living in Eurasia for at least as long as *Homo sapiens* had been living in Africa, and had become adapted to a way of life there so well that today they are viewed as the acme of prehistoric cave life. Neanderthals had no shortage of smarts: They had brains as large as or larger than those of modern humans. For tens of thousands of years, Neanderthal stone-tool-based technology was essentially the same as that of *Homo sapiens*.

The key was population size. It is highly probable, indeed likely, that modern humans lived in slightly denser communities, and that these communities were better connected to one another over longer distances, making them more resilient to localized shocks such as a temporary absence of game to hunt, or food to forage. It's also possible that *Homo sapiens* was slightly more successful at producing children and raising them to adulthood than Neanderthals. The differences need only have been tiny but would have become significant over many generations.

In the end, Neanderthal populations were so small, so widely scattered and so divided by the increasing mass of modern humanity spreading out in between that they could no longer compete, except by assimilation. The same seems to have been the case for the Denisovans of eastern Asia. Other hominins – *Homo floresiensis* and so on – would have become confined to ever smaller refuges until they, too, winked out. Their fate could, of course, have been grislier – as one commentator has noted, *Homo sapiens*' school report might well say that it doesn't play well with other children.

By 40,000 years ago, at the latest, *Homo sapiens* was the last hominin on Earth. In just a few geological eyeblinks, it had not only eliminated all the other hominins but had spread to every part of the Earth.

It had reached that point in its evolution comparable with the greatest extent of the Roman Empire under Trajan, the point at which Gibbon began *The Decline and Fall of the Roman Empire*. After the early second century, the Romans found that they had reached their limits, and any attempts to exceed them were likely to be either short-lived (Mesopotamia, Germany, Romania) or insuperably difficult (the Persian Empire).

The natural range of *Homo sapiens* came to encompass the entire Earth, something that no other species had ever accomplished. And as the Helsinki researchers discovered, when a species succeeds in eliminating the competition, the only way is down.

# PART TWO
# FALL

# 5
# AGRICULTURE: THE FIRST CASUALTY

*From the lines, the galleys, and the bridge, the Ottoman artillery thundered on all sides; and the camp and city, the Greeks and the Turks, were involved in a cloud of smoke, which could only be dispelled by the final deliverance or destruction of the Roman Empire.*

EDWARD GIBBON,
*The Decline and Fall of the Roman Empire*

My friend Brian Clegg is a professional writer. He has written more books than most people have had hot dinners, and so knows a thing or two about writing. The most important tool a writer can have, he says – more than a computer, or a printer, or a handy sheaf of blank paper, or even a pencil – is a dog.

Even Brian gets stuck, you see, and there is nothing better than a dog to tear you away from your desk for some necessary exercise in which you can let your mind roam free, and by the time you get home, you have solved the problem that had inexplicably boxed you into a corner just an hour beforehand.

Writing, too, is a sedentary occupation, often undertaken

in a seated position (bad for posture). Writers need to get up from their desks from time to time and work their muscles. They can always put off going to the gym, but the sight of those appealing doggy eyes and waggy tail, all anticipation for the exciting ball-chasing-and-running-around to come, is irresistible. I've taken Brian's tip but have gone overboard a little – I don't have one dog, but four, and even though I don't usually take out more than three at a time, my daily walks look like I'm fighting a losing battle trying to train a dog sled team. But it gets me away from my desk.

People and dogs have been together for a long time. It's almost certain that dogs were the first of the many animals that humans domesticated. Nobody knows precisely when or where dogs were first domesticated, but it was likely in Europe between 40,000 and 20,000 years ago.[1] Dogs started as wolves that hung around human campsites, just shy of a stone's throw away from the campfire. They were tempted by tasty food scraps and offered services in return such as ridding campsites of vermin, and warning of approaching predators, or other people. It's worth noting that dogs were domesticated during the coldest part of the most recent ice age, so it could have been that dogs and humans teamed up as a way of facing the cold together: Humans would have got companions; dogs, the leftovers from meals too rich in meat protein for humans to assimilate.[2] Dogs and humans are social carnivores and scavengers, with similar social lives, so it was perhaps not too great a stretch to imagine the packs of people and dogs moving as one. When humans struck camp and moved on, they would have been followed – initially at a discreet distance – by a pack of ever more docile and socially habituated wolves. These were the first dog walks, recreated every day by me and millions of other dog walkers during our daily exercise.

When I was a child, I had the good fortune to live in a house with no immediate neighbours, set in what seemed to me an inexhaustible woodland. While walking the dogs (naturally) I got to know which mushrooms grew where, which were edible and which fruits of the forest could be gathered at certain times of year. August and September were times for collecting wild blackberries (*Rubus fruticosus*), but October was a special treat, because that was when my sister and I would go to a grove of sweet chestnut trees (*Castanea sativa*) we knew. We'd eagerly gather up the prickly fruits for the tasty treasures within, which we'd take home so my father could roast them over the fire. Had I been inclined to hunt (I wasn't), I knew which woods tended to be home to deer, where the rabbits had their holes and where to find birds' eggs.

Today, I live in a very different environment – the coast. I know which beaches are best for oysters (*Ostrea edulis*), or razor clams (*Ensis magnus*). Cromer, my adopted hometown, is famous for its edible crabs (*Carcinus pagurus*). I know the places where seals come ashore to breed. I know where to gather samphire (*Salicornia europaea*) on the salt marshes during its all-too-short season in the early summer.

Hunters and gatherers do not, as a rule, roam randomly across a landscape. They tend to follow a predictable path, so that they can best exploit natural food sources as the seasons change. They know that herds of wild game will be in specific places at certain times of the year. In the same way, they will aim to be in the right groves when the fruit is in season, at the right lakes and rivers when the fish are biting, and so on. It was thanks to habits like this that farming started.

The Fertile Crescent is that swathe of land that extends north along the eastern Mediterranean shore from southern Israel to Turkey, and then inland and south-eastwards along the valleys of the Euphrates and Tigris. It was here, more than 10,000 years ago, that people began to settle down. The habits of hunters and gatherers had become so predictable that some of them – the so-called Natufians – began to live in small villages and commute daily to wherever they needed to hunt or gather. As they did so, they began to change the objects of their desire.

Wild grasses bear seed pods which, when mature, shatter, spreading the seeds round about. Early gatherers tended to harvest the grains when they were as mature as they could be, but before the pods shattered, so they could take the pods home and grind them into flour. These gatherers would select those stands of grasses with the biggest, non-shattering ears of grain, and perhaps scatter a few of the grains closer to home – at first by accident, and then deliberately. These people became the first farmers.

That's how the story goes, anyway, but the origins of agriculture are debatable and complex.³ The main question is why, after hundreds of thousands of years, *Homo sapiens* became sedentary and turned to agriculture. It wasn't just in the Fertile Crescent, either (where people first domesticated wheat and barley), but many times independently. Within the space of a few thousand years, farming began in the eastern United States, Mesoamerica, the Andes, Amazonia, tropical west Africa, Ethiopia, China and the highlands of New Guinea. From the perspective of geological time, the invention of agriculture was worldwide and simultaneous.

The answer might lie in a collision of three factors. First, the world was just emerging from its coldest snap for hundreds of thousands of years. 26,000 years ago, glaciers covered much of Europe and North America, but the recovery was far from slow and steady. Before it settled down into the current tolerably warm phase about 10,000 years ago, the climate switched back and forth between warm and cold snaps with a rapidity so marked that it would have been noticeable on the scale of individual human lifetimes. The constant changes in climate were playing havoc with once-dependable food sources, forcing people to seek more reliable means of making a living.

Second, there were more people. After millions of years in which hominins were minor players, human populations were beginning to increase to such an extent that a life of hunting and gathering might no longer have been sustainable. Just how large the world's population would have to have been to reach what in technical terms is called 'carrying capacity' is a matter of debate that I'll return to later. However, it's a fair supposition that the world population of *Homo sapiens* was beginning to strain the limits of what the planet could provide to humans that subsisted solely on hunting and gathering. The human population was beginning to make itself felt in other ways, too.

Third – and something of a consequence of the second – the increasing human tide had depleted the world's wildlife. The interval between 26,000 and 10,000 years ago was marked by the extinction of most animals more massive than about ninety pounds – about the size of a large dog.[4] This was most noticeable in the Americas. There is a great deal of debate about when humans first entered the Americas – presumably over a land bridge from Asia in the far north – but there is increasing evidence that they made landfall around the peak of the last major glaciation, around 26,000 years ago, and possibly earlier.[5] Thereafter they swept through the continent from north to south with great speed, a movement coincident with the extinction of most large animals, including the gigantic ground sloths and armadillos the size of cars. The depletion was almost as marked in Australia, with the demise of giant kangaroos and wombats the size of hippos.

Across Eurasia, iconic species such as the woolly mammoth (*Mammuthus primigenius*), the woolly rhinoceros (*Coelodonta antiquitatis*) and the giant deer (*Megaloceros giganteus*), marked by its immense antlers, became extinct. In these instances, the causes of extinction were more nuanced. Climate change played its part, by diminishing the distinctive ecology of the far north, with its 'mammoth steppe', a kind of environment with a diverse mixture of plant and animal life of a kind not seen today. The effect of human beings thrown into the mix would have been to accelerate an extinction that was all but inevitable. The result, though, was the same, either way. By climate or human agency, hunting large game became unsustainable by virtue of the absence of large game to hunt.

## AGRICULTURE: THE FIRST CASUALTY

Even so, the menu of animal and plant species remained varied, and it would be easy to overplay the immediate effects that people had on wildlife. Another question, raised by Jared Diamond,[6] is why, given this variety, people managed to domesticate so little of it. Even though most large animals vanished, no fewer than 148 species of large animals were left, of which people managed to domesticate just fourteen. We are all familiar with cows and horses, pigs and llamas, goats and sheep, and possibly the Asian elephant, but where are the domesticated zebras, rhinos, hippos and giraffes? As for antelopes, there remains an entire alphabet soup of creatures from eland to nilgai, by way of gnu, topi, kob, dibatag, hartebeest, kudu, pudu, oryx and bongo, none of which are domesticated. Of the 200,000 species of flowering plants, only about one hundred have been domesticated such that they support large numbers of people, and of those, a handful remain dominant – wheat, barley, rice, millet, sorghum and a few others. As Diamond notes, the failure of people to domesticate more than a small fraction of the available wildlife is not for want of trying. The fact is that most creatures are resistant to it. Zebras are notoriously vicious. Bison will not be corralled and can leap six-foot fences from a standing jump. Oak trees – even those few with a mutation that allows acorns to be palatable – have lifespans much greater than that of any individual human, a factor that precludes useful plant breeding.

Increasing pressure of population combined with a reduced variety in diet had a catastrophic effect on human welfare.[7] The trend towards a more sedentary life, and a reliance on a rather small number of starch-rich plant foods, contrived to cause a sharp downturn in human health. The first farmers didn't grow as tall as their hunter-gatherer forebears, and this stunting was most pronounced in childhood and early youth. Malnutrition was widespread. Dependence on only a few crops left populations vulnerable to crop failure caused by inclement weather.[8] Skeletons from early farming communities show a marked uptick in incidence of tooth decay and bone malformations related to dietary shortages of micronutrients such as iron and various other essential minerals – deficiencies absent from hunter-gatherer populations. They also show signs of increased infectious disease.

Not that hunter-gatherer populations were disease-free – they were not. As wild animals do, they laboured under a load of parasites such as worms, fleas and lice.[9] However, they were generally free from the kinds of disease that spread from person to person, because there weren't many people, and they were separated by distance and natural barriers that would have hindered spread. Neither did they contract the kinds of disease that spread between animals and people, because people did not stay long enough in the presence of an animal before the latter was killed, cooked and eaten.

Diseases that spread between people and animals took off when people started to live for long periods of time crammed together in small, dense clusters of permanent dwellings, often along with their domestic animals, such as chickens and pigs. Such proximity, combined with a general absence of sanitation, and a rise in the kinds of vermin that thrive in human company, provided an ideal breeding ground for a whole new spectrum of infectious diseases. The advent of diseases such as tuberculosis (a disease of cattle), plague (spread from rats, via fleas) and influenza (domestic poultry) was a product of agriculture. That respiratory diseases can spread from animals to humans is now, tragically, well known to everyone on the planet – I write this during the continuing SARS-CoV-19 (coronavirus) epidemic. Although the source of this disease has been a matter of controversy, most scientists agree that the virus that causes the disease jumped from animals to humans as a result of direct contact, though the animals concerned might have been wild-caught creatures rather than domestic animals.[10] Domestic poultry, though, are the source of another, less well-publicized pandemic – avian influenza, which is spreading through the world's poultry flocks, and has now started infecting cattle, where it spreads through milk.[11] It is less well-publicized than SARS-CoV-19 because it does not easily spread to humans, but when it does, the result can be fatal.[12]

The strange thing was that agriculture – with its outriders of famine and pestilence – did not cull the human population but increased it further. Hunter-gatherers tend to have children infrequently: a woman usually weans a child before becoming pregnant with another. However, agriculture tended to promote earlier weaning and thus more frequent pregnancy, thus producing more people, more quickly, even compensating for the higher mortality.

Evidence from small, spouted vessels from Bronze Age and Iron Age archaeological sites suggests that infants were fed milk from ruminants, such as cows or sheep.[13] As infants began to drink milk from ruminants, so did adults. And this is odd, because adult humans, until recently, generally lacked the capacity to digest lactose, a sugar that's present in milk. Although many adults today are lactose intolerant, the ability to digest lactose spread through human populations after the advent of agriculture. At first, people drank ruminant milk as a fallback food during times of famine.[14] But dairying soon spread, and nomadic pastoralists milked their flocks (and even their horses) from an early date.[15] The ability to digest milk, even as adults, shows that agriculture had an even more profound effect on humans than simply exposing them to malnutrition and disease. Agriculture made humans evolve.

Such evolution is manifest in another scourge, seemingly of modern times. That is the epidemic of obesity, particularly associated with the incidence of type 2 diabetes. Hunter-gatherers eat a wide variety of foods, but because such people live a marginal existence, never knowing where their next meal will come from, they will tend to gorge themselves on those rare occasions when food is abundant, especially if it is of high calorific value. Even though – as Jared Diamond has elsewhere noted[16] – we do all our foraging in supermarkets and are used to a life of abundance, we still carry an ancestral tendency of eating to excess when food is available. But when food is available all the time, especially foods rich in starch and sugars (as agricultural products generally are, compared with their wild relatives), then we run the risk of developing what scientists call 'metabolic syndrome', in which an impaired tolerance to high concentrations of glucose in the blood may lead to type 2 diabetes.[17] In modern societies, though, diabetes is much less prevalent among people of European ancestry than among, say, Pacific Islanders. Diamond sees this as a product of evolution. Because Europeans transitioned to a starchy, sugar-rich diet a long time ago, the epidemic of diabetes in Europeans has already played out. People with a genetic predisposition to diabetes tended to die without producing offspring, their genes perishing along with them. But the epidemic is still with us, especially in people who transitioned to the Western diet more recently.

This evolution can be seen even today. One of the highest incidences of metabolic syndrome in the world is in the population of the small Pacific Island nation of Nauru. According to a study of people on the island, in 1987 almost a quarter (24 per cent) of Nauruans had type 2 diabetes. However, by that date, the incidence of glucose intolerance (a precursor to diabetes) had declined from 21.1 per cent in 1975–6 to 8.7 per cent. The authors of this study[18] note that as people with type 2 diabetes tend to have higher mortality and lower fertility than those who are not diabetic, it could be that those Nauruans with a tendency to diabetes are being weeded out of the gene pool – something that happened in Europe centuries before. Even 10,000 years after its advent, the consequences of the transition to high-calorie, high-sugar diets continue to be felt.

An increasing population and dependence on a relatively small number of crops exposes people to famine. Before the Industrial Revolution, people in Europe experienced famine regularly, mostly as a result of bad weather, though famines were most prevalent when populations were high. It's notable that they were rare in the two centuries following the Black Death of the fourteenth century. That pandemic presumably killed off so many people that the survivors enjoyed a period of relative surplus.[19]

After the Industrial Revolution, famines became rare, at least in Europe, though they did not cease entirely. Latter-day famines tend to be caused, or, at least, made worse, less by nature than by tyranny or mismanagement. One thinks of the early communist Soviet Union, or the Great Leap Forward in China. A particularly noteworthy example was the Irish Potato Famine of the 1840s, caused by the infection of potatoes with the fungus-like pathogen *Phytophthora infestans*, the cause of potato blight. Although this disease raged through Europe at this time – and is still around today – the effects in Ireland were exacerbated by bad weather; the overdependence of a large population on a single staple crop; spectacularly bad to non-existent land management by absentee British landlords; and lack of sympathy from the government in London. Millions died from hunger; more millions emigrated, especially to the United States, leading to a dramatic decline in the homeland population.

There should be a lesson here for people today. Food security is challenging to maintain given that large populations of people are dependent on a small number of crops, especially in a world in which those large populations are increasingly interdependent. I write this during the Russian invasion and occupation of large areas of Ukraine, a nation that produces a significant amount of the world's wheat. The war has disrupted the growing and distribution of the crop, leading to a dramatic rise in food prices worldwide.

Lessons, though, are never learned quickly. Not only do farmers grow a small number of crops, but those crops are limited to strains or breeds that yield large amounts of produce when farmed intensively. Consider, for example, the banana.[20] This plant was domesticated in south-east Asia some 7,000 years ago, but half the current worldwide production relies on a single strain, the Cavendish. Worse still, all Cavendish bananas are clones – which means they are genetically identical. This has made it a magnet for various pests and diseases, offering a grave threat to the production of this crop. Now, the world could probably live without bananas. But the lesson is clear for all crops.

Hunter-gatherers exist in a state teetering on the edge of starvation, but they do not, as a rule, suffer from the wholesale malnutrition and other health problems consequent on the invention of agriculture, with its additional burden of increasing population density, disease and famine. They also subsist on a much wider variety of foodstuffs than post-agricultural populations, with their dependence on a relatively small number of starchy staple crops that leaves people exposed to famine when those crops fail.

Given the obvious benefits of agriculture, you might complain that I protest too much. To be sure, billions of people still lack access to clean water and a balanced diet, but if they do, this absence has more to do with such human foibles as poor governance, corruption and warfare than crop failure per se. People are (in very broad terms) better governed, more educated, better nourished and less inclined to conflict than they were even a few centuries ago. Or are they? According to the International Monetary Fund, food insecurity has been rising since 2018 (even before the invasion of the grainlands of Ukraine by Russia), a consequence of the increasing frequency of climate shocks such as flooding, storms and drought – a consequence of climate change – exacerbated by regional conflicts.[21] In 2023, some 238 million people in forty-eight countries are facing what ReliefWeb[22] calls 'high levels of food insecurity', 10 per cent more than in 2022.

Famine is riding down hard on us, faster than ever. This comes at an inauspicious time, for never in human history have so many people been dependent for their calories on so few crops. It isn't just the crops that have been depleted in the genetic resources they need to face increasingly uncertain times. Humans are running out of genetic resilience, too.

# 6

# POX-RIDDEN, WORM-EATEN AND LOUSY

*The triple scourge of war, pestilence, and famine afflicted the subjects of Justinian, and his reign is disgraced by a visible decrease of the human species, which has never been repaired in some of the fairest countries of the globe.*

EDWARD GIBBON,
*The Decline and Fall of the Roman Empire*

Cast your mind back to 1652. That was when the Dutch East India Company set up a small colony at the Cape of Good Hope, on the southernmost tip of Africa. The reason for establishing such a far-flung outpost was to provision ships sailing round the Cape for the long onward journey to what was then the East Indies. Some of the settlers left the company and started farming on their own account. As is so often the case with farming communities, there was a deficiency of women: The farmers were in want of wives. To fulfil this need, the company sent out a shipment of orphan girls in the hope that they would marry the farmers.[1] So it was that in 1688, a farmer named Gerrit Jansz van Deventer wed one Adriaantje Adriaanse van Rotterdam.

But there was a flaw. Either Gerrit or Adriaantje carried a genetic mutation predisposing to a disease called porphyria variegata. In this condition, the body has problems breaking down excess haem – the iron-containing substance that carries oxygen round the bloodstream and makes it red. The symptoms are varied and include discoloration of the skin, especially when exposed to sunlight (a circumstance that has led to lurid claims that porphyria is a real-life explanation for the aversion of mythical vampires to daylight).

Each parent contributes two copies of any given gene to their child. Usually, a healthy copy of a gene from one parent will cover for the mutant copy from the other, and the child will not suffer from the disease (though they can still pass it on to their own children). Porphyria, however, is different. A mutant copy of the faulty gene from either parent can lead to disease in a child. The disease can affect boys and girls, though the symptoms can vary hugely. Some patients have barely any symptoms at all: The presence of porphyria in their family didn't stop Adriaantje and Gerrit founding a dynasty. A fair proportion of Afrikaners (South Africans of Dutch descent) can trace their ancestry to this union – perhaps 40,000 South Africans today have inherited the mutation, so much so that porphyria variegata, of all the different kinds of porphyria that exist, is especially associated with South Africa.

The incidence of porphyria variegata in South Africa is much greater than in the human population at large. It is certainly much greater than its incidence in the Netherlands, where both Adriaantje and Gerrit's ancestors came from. However, the Cape colonists were a small subset of the

Dutch population more generally, and because just one of the colonists had the faulty gene, the incidence of the disease in the colony was proportionally much larger. The skewed incidence of porphyria variegata in Afrikaners is just one example of what is called the 'founder effect'.

The founder effect works like this. Imagine you have a bag of one hundred rubber balls. You know that ninety-nine are blue and one is red. So, the proportion of red balls is one out of a hundred, or 1 per cent. Now, imagine taking ten balls randomly out of the bag. Most of the time all the balls you select will be blue. However, there is a small chance that the red ball will be among the ten. Even though the number of red balls hasn't changed, the proportion of red balls has leapt from one in a hundred (in the bag as a whole) to one in ten (in the sample you've picked). This explains how a mutation (the one red ball among all the blue ones) that was rare in the Dutch population happened to become much more common among the small sample of Dutch founders of the Cape Colony.

In the early days of the Cape Colony, Dutch colonists had very few choices of mate, and most of the potential pool of partners will have been related to one another. This increases inbreeding, and one of the effects of inbreeding is to amplify the incidence of otherwise rare mutations. The porphyria mutation, already hugely amplified by the founder effect, spread still further. The founder effect, followed by inbreeding – necessarily so, given that founder populations are small – is common in small societies in which finding partners outside the group is discouraged. Such societies tend to be prone to particular disorders. Bipolar disorder is more common among the Pennsylvania

Amish[2] than in the population at large, and the same is true for the incidence of Crohn's disease among Ashkenazi Jews, that is, Jewish people of primarily eastern European origin.[3] We have already met the highly inbred Habsburg dynasty of early modern Europe – a society within a society – and sympathized with poor benighted Charles II, final scion of the Spanish branch.

In some cases, genetic diseases that afflict some populations more than others can have an advantage – a relatively mild genetic disorder can protect a sufferer from something worse. In addition to Crohn's disease, Ashkenazi Jews suffer disproportionately from Gaucher disease, an inherited disease that can lead to swollen internal organs (principally the spleen and liver) and a propensity to bone fractures. Symptoms can vary from severe to sufferers hardly being aware of the disease. However, there is some suggestion that people with the disease are relatively immune to tuberculosis. Substances that accumulate in the cells of people with Gaucher disease are potent agents against *Mycobacterium tuberculosis*, the bacterium that causes tuberculosis. In the Middle Ages, Jewish populations in cities were invariably required to live in crowded ghettos, where diseases such as tuberculosis would have been rife. It is possible that natural selection allowed the otherwise harmful Gaucher disease mutation to thrive because of its beneficial effects in warding off tuberculosis.[4]

It seems that the more one looks for it, the more the founder effect turns up in human populations, ancient and modern, from Finns to Roma, from the residents of Papua New Guinea to French Canadians. Genetic analysis shows that many modern groups of people, from Native Americans to Pacific Islanders, have experienced founder events in their history even more marked than that in Ashkenazi Jews, because of either a modern population that descended from a very small group of founders, or a historically low population size.[5] The founder effect is, perhaps unsurprisingly, particularly marked on islands, where populations are always small. It turns up in islanders from the Orkneys to Malta, Sardinia to Iceland. Founder effects are also strong among hunter-gatherers such as Pygmies; nomads such as Bedouin; and indigenous and tribal groups in general. Given that all humans were hunter-gatherers for almost the entirety of their history, this can only be seen as significant. For virtually the whole of human history, *Homo sapiens* existed in tiny groups, many of which died out, or, if they survived, did so thanks to a few hardy souls that clung on in the face of extinction and founded the next generation.

The lesson of the founder effect is clear. Populations that all descend from a small number of founders will have little genetic variation. They'll tend to have the same ailments, the same predispositions to the same diseases. The same seems to have been true of the entire human species. The apparent differences in obvious, visible traits such as skin pigmentation between people sampled from all over the world are, quite literally, skin deep. As noted before, there is more genetic variation in one small population of wild chimpanzees than in the entire population of *Homo sapiens*. And it's not just chimps: Populations of gorillas and orangutans may be small, endangered even, but are much more genetically varied than humans.[6] The implication is that the human species has been through what's called a 'bottleneck'. In such an event, most of the species will have become extinct, extinguishing almost all the genetic variation. The continuation of the species will depend on a small and perhaps random sample of those that are left – the founder effect, but for the whole species.

It's increasingly clear that the whole of humanity outside Africa descends from a small population that left that continent sometime between 100,000 and 50,000 years ago. This is shown by the fact that human diversity in Africa is greater than in the rest of the world and has a deeper history. In other words, the totality of non-African humanity represents just a small twig on a large and sprawling African bush.

But there is more to it than that. Human diversity is small, relative to that of chimps, gorillas and orangutans even when African diversity is taken into account. This means that the founder effect responsible for the relative deficiency of human diversity must run back to some bottleneck, some pinch point, in the days before humans left Africa.

In my book *A (Very) Short History of Life on Earth*, I recount a story of the near extinction of *Homo sapiens* even before our species left Africa. Never common, the small bands that constituted our species winked out first here, and then there, until the entire species was confined to an oasis in the Kalahari Desert. There this tiny remnant stayed for 70,000 years, until a spell of more clement climate allowed it to break out of its confines, spread throughout Africa, and then, the world. If ever there were a Garden of Eden, it would have been at this oasis, the once lush Lake

Makgadikgadi, which has since dried out to become an inhospitable salt pan.[7]

It makes a good story, but it has not found general favour in the scientific community. The current consensus is that the African genesis of *Homo sapiens* involved several populations spread throughout Africa, ever diverging and mixing in a loose coalition over tens – even hundreds – of thousands of years.[8] I do believe, however, that it is true – even if only poetically. Whether or not Makgadikgadi was the birthplace of all humans alive today, genetics tells us that *Homo sapiens* was driven almost to extinction in the past, not just once but possibly several times. More recent work suggests that even further back in human prehistory, the ancestors of modern humans living in Africa declined to a population of no more than 1,280 breeding individuals, and they remained like that, teetering on the edge of extinction, for more than a hundred millennia, between about 930,000 and 813,000 years ago.[9]

Because every human on the planet is descended from a small group of people that lived in Africa long ago, the founder effect is extended to an entire species. And, just like with the founders of the Cape Colony, *Homo sapiens*, perhaps by reason of this founder effect, is prone to an extraordinary litany of ailments.

Despite their hideous variety, infectious diseases fall into just six classes.

First, there are the worms – parasites such as tapeworms, roundworms and the fluke *Schistosoma* that causes bilharzia, scourge of the tropics.

Then come the protists, single-celled organisms such as the malarial parasite, the trypanosomes that cause sleeping sickness, and agents of dysentery such as amoebae and *Giardia*.

Next up are the bacteria – single-celled organisms much smaller and simpler in structure than protists, though vastly more numerous. Bacterial diseases include ancient afflictions such as the plague bacillus *Yersinia pestis*; the agents of tuberculosis and leprosy; venereal diseases such as gonorrhoea and syphilis; diphtheria, anthrax, typhoid, tetanus and cholera; virulent strains of usually harmless *Staphylococcus aureus* (a common bacterium that lives on human skin and can become a pestilential problem in hospitals); and *Escherichia coli* (a common bacterium that lives in our guts and can occasionally go rogue).

After that there are fungi – a relatively minor source of complaints – such as athlete's foot and thrush, although some fungal infections, such as aspergillosis (which occurs when the fungus *Aspergillus* gets into the lungs), can be fatal if left untreated.

Fifth, there are viruses, such as measles, chickenpox, rabies, smallpox, mumps, rubella, poliomyelitis, yellow fever, influenza, HIV-1, Ebola, SARS-CoV-19 and, that eternal irritant, the common cold. Viruses, as a rule even smaller than bacteria, are tiny packages of genetic material that remain dormant until they gain access to much larger cells, which they subvert to become factories for making more viruses. Viruses don't just infect humans and other animals. They attack protists, bacteria and even other viruses.

A sixth category includes a small but mysterious and lethal class of brain diseases caused by agents called prions.

These are misshapen versions of normal proteins in the body that transmit their rogue forms to other normal proteins, thereby causing molecular havoc. Such diseases include Creutzfeldt–Jakob disease, a variant of which is associated with the cattle disease bovine spongiform encephalopathy, or BSE.[10] Prion diseases are mercifully rare, but in one traditional society in New Guinea, a prion disease called kuru[11] was spread by cannibalism practised as a part of funerary ritual.

In his book *Plagues Upon the Earth*,[12] historian of disease Kyle Harper notes that our closest relatives, chimpanzees, suffer relatively few diseases, despite their habit of eating raw monkey meat and their unsavoury practices mentioned in the prologue. We humans, in contrast, labour under the burden of an enormous number of diseases.

Nevertheless, infectious disease is a much greater burden on chimpanzees than on humans, who over the past few centuries have largely eliminated infectious disease as a leading cause of illness or death. Because of advances in sanitation and modern medicines such as antibiotics, humans today are more likely to fall ill from diseases associated with ageing (Alzheimer's, Parkinsonism), lifestyle (cancer, heart disease, stroke), rare genetic (that is, inherited) complaints, or accidents. For all that, we still attract infectious diseases like dogs attract fleas. Kyle Harper offers a checklist of 233 known infectious human diseases, comprising eighty-five viruses, seventy-three types of bacteria, twenty-one protists, seven fungi and forty-seven types of worms (though he doesn't mention prions).

Chimps are known to host around twenty-eight viruses – many fewer than humans – and some of those they have caught from humans. In 1999, almost one in five chimpanzees in the Taï rainforest in West Africa died from a combination of diseases contracted from humans. They were first struck with respiratory syncytial virus (a leading cause of respiratory infection in human infants) and then felled by pneumonia caused by the common bacterium *Streptococcus pneumoniae*.

Worms and protists are more important agents of disease in chimps than bacteria and viruses, just as they were for our hominin ancestors. Our association with worms has increased ever since we came down from the trees and started rootling around in soil (notoriously full of worms, and not just earthworms) in the cause of agriculture and acquiring food and water from places close to where we deposited our own and animal wastes. Humans have acquired diseases, then, as a result of our distinctive history: Viruses and bacteria thrive among humans that live in large, dense and unsanitary groups. Cities – that unique innovation of humans – were, historically, deathtraps. They could only persist where there were sufficient immigrants to replace the urban population that died from infectious disease. It has only been since the invention of modern sewage systems that cities could grow, to become net sources of surplus population rather than fetid sinks of squalid and unsanitary death.

Many human diseases have spread from the animals we have domesticated and hunted, or from those that otherwise have come into proximity with humans. Such diseases, called zoonoses, go right back to the dawn of agriculture. Of all the 233 infectious diseases of humans listed by Kyle Harper, one hundred are zoonoses. The various types of influenza (from birds, pigs and cattle), the plague bacillus *Yersinia pestis* (from rat fleas) and the tapeworm *Taenia solium* (from infected, undercooked pork) are just a few of this lethal legion.

Harper's list understates the problem, however. Although we might contract any one of a hundred different diseases directly from animals, some diseases that are *only* known in humans are closely related to diseases in animals, suggesting that such diseases originated in animals with which humans came into close contact at some time in history. Examples include tuberculosis (from cattle). More recent ones include HIV-1, which is closely related to viruses in monkeys and apes; and, of course, SARS-CoV-19, which is similar to diseases found in bats, known to carry a wide range of viruses.[13] The more that expanding human populations encounter previously undisturbed wildlife, the more likely that some seemingly unlikely animal pathogen will make its way into the human population, finding there a rich and unprotected field of potential conquest.

Many organisms associated with diseases have evolved with us, and some are even unique to the human ecosystem. Perhaps the earliest known is the above-mentioned human body louse.[14]

To be sure, it is the changes in our lifestyle, especially since the invention of agriculture, and the habit of living in large, sedentary groups, that have been the cause of many diseases in humans. But even so, the toll of disease in humans seems excessive. Perhaps it is the genetic homogeneity of humans that helps make us so vulnerable.

Genetic variation is a vital hedge against the spread of disease. More than two billion years ago, living things evolved a way to shuffle the card deck of genetic material, maintaining genetic variation to keep one step ahead of disease. This shuffling mechanism is called 'sex'. In general, living things that reproduce simply by making copies of themselves are more exposed to invasion from disease than those that reproduce sexually. One of the body's systems that depends on genetic variation is the immune system. Most living things have a baseline level of immunity against disease – so-called innate immunity.

However, in some animals (including humans) innate immunity is augmented by another system by which the body can learn from its experiences. If a living creature catches a disease and survives, its immune system retains a memory of that infection, making it better equipped to deal with the threat next time. This is called 'adaptive' immunity and is the basis of vaccination. In vaccination, a person is dosed with a killed or weakened version of a disease organism as a way of 'training' the immune system to deal with the real thing, should it arrive. Currently, vaccination is the only sure-fire way to combat viral diseases.[15] It has been so successful that dreadful diseases such as smallpox and poliomyelitis are extinct, or at least very rare. When I was a child more than half a century ago, measles was a disease that killed or disabled young children. But that was then, and effective vaccines against measles and other childhood diseases such as mumps and rubella now exist. That people seem to have forgotten the terrible effects of measles might in part explain the woeful fact that measles vaccination rates have fallen in some places as a result of misinformation about the relative safety of measles vaccines.

So much for well-known diseases. The major threat to human populations comes from diseases with which people have no experience – usually something that spreads from another animal host. Newly arrived diseases might be especially virulent. That is, the symptoms of disease are extremely serious, life-threatening even. Diseases that stay virulent cut a swathe through the population but usually burn themselves out. Once they have killed all their potential hosts, they themselves become extinct.

History is littered with accounts of virulent diseases that have no known modern counterpart. One such was the so-called English sweating sickness that appeared suddenly in England in 1485, at the end of the Wars of the Roses and coincident with the coronation of King Henry VII, and could kill those infected within a few hours. The last recorded outbreak was in 1551, after which it vanished. The cause of this dreadful disease remains unknown.[16]

Usually, the host and the disease reach some accommodation. The disease becomes less virulent, so that people with it might get sick but usually recover – not before, however, passing the disease on to others. At this stage the disease becomes naturalized as a human affliction.

Before universal air travel, diseases could be naturalized in some populations and yet absent from others. When Europeans first came to the Americas in large numbers in the late fifteenth century, the diseases they brought with them – such as influenza and smallpox – devastated the indigenous populations far more effectively than the conventional instruments of conquest.[17] So much so that the wavefront of disease spread even more quickly than the invaders themselves. When Europeans arrived in many once-populous parts of the Americas, they found them all but deserted. The Europeans might have brought syphilis, too, though it's possible that this disease migrated in the other direction – back to Europe with the returning conquistadores.

Before the invention of modern medicine, humans were prey to plagues and pestilences that could wipe out sizeable fractions of the population. One thinks of the plague that struck the Roman Empire in the time of the Emperor Justinian; the Black Death of the fourteenth century; and the so-called Spanish influenza epidemic that immediately followed the First World War. As of April 2023, SARS-CoV-19 has killed almost seven million people worldwide,[18] its progress aided by air travel. This may seem small as a proportion of the total human population (less than one death per thousand). However, the epidemic would have carried off many more had it not been for the heroic efforts of scientists and clinicians to create effective vaccines.

And yet the fight goes on, and one is entitled to worry about the effects of the next pandemic, which – like English sweating sickness – might come from nowhere and exact a sudden and dreadful toll. New diseases are forever cropping up. Ebola,[19] Legionnaires' disease,[20] Zika,[21] Chikungunya[22] and West Nile virus[23] are just the first five that I could immediately think of without trying, all of which turned up in the twentieth century and have since become causes for concern. As the recent coronavirus pandemic has shown, governments are not always as prepared as they perhaps ought to be for their sudden emergence.[24]

For the fact remains that as a species, humans are remarkably pox-ridden, worm-eaten and lousy, a consequence, at least in part, of a singular lack of genetic variation, the result of multiple founder effects in human history.[25]

Looking back, it could be that human populations have been kept small, and thus prone to founder effects, partly *because* of contagion. Infectious diseases contracted over the history of *Homo sapiens* have left indelible scars on the genes of every human now living,[26] a sign of severe, episodic culls in the past.

The human population is currently greater than it has ever been, with an ever-greater proportion of people living close together in cities, which attract an ever-greater proportion of people on the move because of conflict or climate change. At the same time, human populations are encroaching on those few wildlife habitats that still exist, exposing people to new diseases otherwise seen only in wild animals. It is in such conditions that infectious diseases are likely to thrive.

# 7

# ON THE BRINK

*Twenty-two acknowledged concubines, and a library of sixty-two thousand volumes, attested the variety of his inclinations, and from the productions which he left behind him, it appears that the former as well as the latter were designed for use rather than ostentation.*

EDWARD GIBBON,
*The Decline and Fall of the Roman Empire*

Despite agriculture – and, in fact, because of it – and despite genetic deficiency, and despite disease, the growth in the human population has been inexorable. For almost the entirety of its history, *Homo sapiens* scratched a bare living from what the wild could provide. Such a lifestyle allows for no more than ten million or so humans[1] to live on the entire planet. Farming increased the Earth's 'carrying capacity' to support humans and, with that, the population, both in numbers and in density. Neither genetic deficiency nor susceptibility to disease has been able to hold back the human tide. Indeed, one is entitled to wonder how many more people the Earth would now be asked to support had humans been of a more varied cast, or less susceptible to infectious disease.

The overwhelming wash of humanity made a strong impression on a young biologist from Stanford University on a night-time taxi ride through the streets of Delhi, India, in 1966.

The streets seemed alive with people. People eating, people washing, people sleeping. People visiting, arguing, and screaming. People thrusting their hands through the taxi window, begging. People defecating and urinating. People clinging to buses. People herding animals. People, people, people.

The name of the biologist was Paul Ehrlich, the recollection on page 15 of his 1968 book *The Population Bomb*.[2] Ehrlich took his experience of the slums of Delhi and, in an at times highly charged polemic, extrapolated it to a terrifying future for the whole of humanity. In it he accuses the US scientific establishment, especially biomedical science, of being more interested in lowering death rates than controlling birth rates:

> The establishment in American biology consists primarily of death-controllers: those interested in intervening in population processes only by lowering death rates. They have neither the background nor the inclination to understand the problem.[3]

Ehrlich compounds his criticism with what seem to be draconian solutions:

> One plan often mentioned involves the addition of temporary sterilants to water supplies or staple food.

Doses of the antidote would be carefully rationed by the government to produce the desired population size. Those of you who are appalled at such a suggestion can rest easy. The option isn't even open to us, thanks to the criminal inadequacy of biomedical research in this area. If the choice now is either such additives or catastrophe, we shall have catastrophe.[4]

His geopolitical solutions to prioritizing international aid – which involve the carving up of nation states in so-called underdeveloped countries (UDCs) – read, nowadays, as severe, paternalistic and even naïve:

> Perhaps we should support secessionist movements in UDCs [underdeveloped countries] when the group departing is better developed than the previous political unit as a whole. Perhaps we should have supported Katanga, not the Congo. Perhaps we should now support Biafra, not Nigeria. West Pakistan might receive aid, but not East Pakistan [that is, Bangladesh]. It might be to our advantage to have some UDCs more divided or even rearranged, especially along economic axes.[5]

You can look up the civil conflicts mentioned – the secessionist movements to separate the province of Katanga from what is now the Democratic Republic of Congo (1960–1963), the region of Biafra from Nigeria (1967–1970), and Bangladesh from Pakistan (1971), and wonder at Ehrlich's glossing over of the human cost of these bloody wars, presumably for the Greater Good.[6]

Not that some of Ehrlich's more realistic prescriptions haven't come to pass:

> We need a federal law guaranteeing the right of *any* woman to have an abortion if it is approved by a physician.[7] [original emphasis]

In the United Kingdom, the Abortion Act of 1967 permitted abortion in certain circumstances, but the case of *Roe v. Wade*, the decision by the US Supreme Court guaranteeing the right of abortion, was made in 1973, five years after the publication of *The Population Bomb* – a guarantee that has lately been withdrawn, at least as a federal law.

*The Population Bomb* closes in a desperate, if not sarcastic mood:

> Fortunately, people can be produced in vast quantities by unskilled labour who enjoy their work. In about 500 years, with the proper encouragement of reproduction, the Earth could be populated to a density of about 100 individuals per square foot of surface (land and sea). That is a density that should please the loneliest person.[8]

*The Population Bomb* had a galvanizing effect at the time on people concerned about the effect of overpopulation on the environment (although Rachel Carson's book *Silent Spring* had been published in 1962, making Ehrlich's book a relative latecomer to the party). According to Charles C. Mann, writing in *Smithsonian Magazine* in 2018 – part of a look back at the events of 1968, half a century on – *The*

*Population Bomb* contributed, directly or indirectly, to programmes to reduce fertility in poor places that verge on officially sanctioned cruelty:

> Some population-control programs pressured women to use only certain officially mandated contraceptives. In Egypt, Tunisia, Pakistan, South Korea and Taiwan, health workers' salaries were, in a system that invited abuse, dictated by the number of IUDs [intrauterine devices, a mechanical means of contraception] they inserted into women. In the Philippines, birth-control pills were literally pitched out of helicopters hovering over remote villages. Millions of people were sterilized, often coercively, sometimes illegally, frequently in unsafe conditions, in Mexico, Bolivia, Peru, Indonesia and Bangladesh.[9]

One thinks further of the one-child policy in China, and the efforts of the government of India under the late Prime Minister Indira Gandhi to coerce people to become sterilized. As a joke doing the rounds at the time had it, Indira Gandhi doesn't only tie men in knots, she ties knots in men.

Mann's image of the helicopter is apposite. Apart from China and India, that took it upon themselves to adopt forcible or coercive methods of birth control, one has the distinct sense that methods of birth control were helicoptered in by Western nations, foisting their principles on what have been patronizingly called developing countries, whether they were wanted or not. In the same article, Mann points out that at the time of Ehrlich's fateful Delhi cab ride, the population of that city stood at 2.8 million.

By comparison, the 1966 population of Paris was about 8 million. No matter how carefully one searches through archives, it is not easy to find expressions of alarm about how the Champs-Élysées was 'alive with people'. Instead, Paris in 1966 was an emblem of elegance and sophistication.[10]

Although the population of the world has more than doubled since *The Population Bomb* was published (just over eight billion at the time of writing this book), Ehrlich's dire predictions of worldwide catastrophe have, thankfully, not come to pass. In broad terms, people around the world are living longer, healthier, safer lives than they did in 1968, without having the water supply laced with contraceptives, or Ehrlich's more extreme geopolitical solutions having been enacted. They are also better educated – and it is from education, especially of women, whence all these other blessings flow. Across the world, the proportion of people in their late twenties with at least six years of schooling rose from just over half in 1970 to more than four fifths in 2018 and is expected to be close to nine out of ten by 2030 – though progress in secondary and tertiary education has been slower.[11]

*The Population Bomb* is perhaps best seen as a product of its time. 1968 was a febrile year. That was the year NASA launched the Apollo 7 and Apollo 8 missions to the Moon, starting the first, brief flirtation of humanity with manned space exploration. It was a year of student revolution. It was also the year of the Prague Spring – political liberalization in Czechoslovakia, brutally repressed by the Soviet Union.[12] It was the year of protests against the Vietnam

War. It was the year in which both Dr Martin Luther King, Jr, and Senator Robert Kennedy were assassinated.

One event of the so-called Swinging Sixties generally goes unremembered. It was in the 1960s that the rate of increase of the human population, trending ever upwards since prehistoric times, reached its peak. According to United Nations data,[13] the rate of population growth peaked in 1964, at 2.24 per cent a year, when the world population stood at 3.267 billion. No wonder Ehrlich felt moved to write his incendiary tract. Since then, the growth rate has slumped to 0.88 per cent and is projected to become negative – that is, the population will start to shrink – in 2086, when the world population will top out at 10.431 billion.

This slowdown in the rate of population growth is perhaps the most important event to occur in the 1960s. One could argue (and, indeed, I do so here) that it is an event in human history as important as the invention of agriculture. At no time in the human career since the advent of farming, except during world wars and pandemics, has the rate of increase of the whole species started to slow down. It is an event that is unique in the entire course of human evolution. The implication is that the rate will soon equal zero (when the birth rate and the death rate will be equal), after which the population will start to shrink. It will appear to do this of its own accord, with help required neither from geopolitical tinkering, forced sterilization or airdrops of contraceptives, nor from pandemics, global wars, climate change, rampaging artificially intelligent killer robots, the invasion of malevolent aliens, or other external agents, although such things might make the decline happen faster.[14] And it is likely to do this soon.

Prediction is notoriously difficult, especially about the future.[15] Or, as Christopher J. L. Murray of the Institute for Health Metrics and Evaluation at the University of Washington, Seattle, and colleagues put it:

> A fundamental challenge in making long-range forecasts, regardless of the modelling strategy, is the existence of potential changes in trends far in the future that cannot be predicted.[16]

Not that they don't have a jolly good try. In a sophisticated and detailed analysis of population trends, Murray and colleagues suggest that the turnround in global population could be even more stark, topping out in 2064 – two decades earlier than the UN projection – at a population of 9.73 billion, declining to 8.79 billion by 2100. However, if UN Sustainable Development Goals (SDGs) for education and contraceptive need were achieved, the population in 2100 could be as low as 6.29 billion – much lower than it is now. It should be said that these figures are hedged around with considerable uncertainty, and Murray and colleagues express doubt that the SDGs will all be met.[17] But the message is clear – for the first time in history, the human population is on a marked downward trend.

It's worth unpacking Murray and colleagues' study, for it is rich in telling detail (it's also freely accessible for anyone to read – I encourage you to do so). Not only that, but it is an important analysis of an important moment in human evolution – the time when the growth of the human population is swinging dramatically into reverse.

One of the key concepts in the study is what demographers – people who study population trends – call the total fertility rate (TFR). When boy meets girl and they settle down and have two children, they are doing no more than replacing themselves in the population. Two parents, two children. If they have more than two children, the population will increase. If they have fewer, the population will decrease. TFR is usually expressed as the average number of children one woman will have in her lifetime. To keep the population stable, TFR must have a value of 2.0. In actual fact, it's slightly more than 2.0 – more like 2.1 – to accommodate various factors, such as children dying in infancy, and also that slightly more male babies tend to be born than female ones.[18] A TFR value of about 2.1 is known as the replacement level. TFR is strongly related to educational attainment and, not surprisingly, supply of contraception. The more that women are educated and have access to contraception, the lower the TFR.

In this context, Murray and colleagues' headline results make eye-catching reading. Of 195 countries and territories studied, 151 are forecast to have a TFR lower than the replacement level by 2050, and they'll be joined by thirty-two more by the end of the century. Furthermore, global TFR will dip below replacement level in 2034.

For some countries the problem is already acute. Twenty-three countries are forecast to lose more than half their populations by 2100. It's no surprise that Japan is among that number, a country well known to have a rapidly ageing population, low birth rate and little net immigration. It's perhaps more of a surprise that countries such as Spain and Thailand are already in this select group. One might have suspected that these countries are starting to show signs of an ageing population and a falling birthrate, but that their populations will halve in less than a century comes as a shock. Lying just outside the cut-by-half club, with a projected net loss of population of 48 per cent by 2100, is the behemoth that is the People's Republic of China. With a TFR of 1.53, China's population of 1.412 billion (in 2017) was predicted to peak at 1.432 billion as soon as 2024 and decline to 732 million by 2100. Prediction being as hard as it proverbially is, these figures are hedged by wide margins of error. The population of China in 2100 could be as high as 1.499 billion (higher than today) or as low as 456 million. But it must count for something that as early as 2022, it was reported that more people died (10.41 million) than were born (9.56 million) in China for the first time since the disastrous famines associated with Mao's Great Leap Forward in the 1960s.[19]

The handful of countries for which the rate of reproduction will still exceed the replacement rate in 2100 include African nations such as Chad (TFR projected to be 2.19 in 2100), Somalia (2.57), South Sudan (2.46) and Zimbabwe (2.22); two Central Asian nations, Kyrgyzstan and Tajikistan (both 2.25); and a cluster of small island nations in the Pacific including American Samoa (2.13), Guam (2.26), Kiribati (2.32), Samoa (4.47) and Tonga (2.62). In most areas, there will be fewer humans in 2100 than now – only in sub-Saharan Africa, North Africa and the Middle East will there be more.

For context, the population of the United States was 324.84 million in 2017. It will peak at 363.75 million in 2062 before declining to 335.8 million in 2100. The TFR in 2017 was 1.81 – well below replacement – but will fall still further to 1.53 in 2100. The population of the United Kingdom was 66.64 million in 2017 and will peak at 74.87 million in the same year as the United States – 2062 – before subsiding to 71.45 million in 2100. The TFR was 1.73 in 2017 and will slide to 1.61 in 2100.

A surprise is the small but extraordinarily fecund nation of Israel, with a projected TFR of 2.36 in 2100, down from 2.9 in 2017. This means that the tiny country will swell from 8.95 million in 2017 to an astonishing peak of 24.07 million in 2100 before falling. These circumstances are due to a high level of immigration, compounded by the large family sizes of an increasing population of Haredi, that is, ultra-Orthodox Jews. When I first visited Israel as a student in 1985, the sight of ultra-Orthodox Jews, with their distinctive garb, was a comparative rarity. On my return in 2017, thirty-two years later, it was entirely normal.

Perhaps more of a surprise is that the rates of reproduction in nearly all countries in sub-Saharan Africa are projected to fall below replacement by the end of the century. At the moment, however, most of the world's countries where the TFR is still well above replacement are in that subcontinent. This is because the populations in those countries are still younger than average, with a greater proportion of the population still of childbearing age. Murray and colleagues predict that fertility in sub-Saharan Africa will not fall below replacement level until 2063, almost thirty years (more than a generation) after the world in general. Even in sub-Saharan Africa, though, populations will eventually peak and start to fall. The most populous nation in Africa is Nigeria. It is already home to more than

200 million people and is forecast to peak at 791 million in the last year of the twenty-first century before it, too, starts to fall. The TFR in 2017 was 5.11 – by the end of the century it is projected to be 1.69. Even so, Nigeria will be the second most populous country in the world in 2100.

In the meantime, though, the world's population is still increasing. Because the rate of increase is uneven, and because it will take some time for the dip in TFR in countries in sub-Saharan Africa to take effect, the make-up of the global population will change. Over the coming decades, the global population will become more African. This will be moderated somewhat by changes in life expectancy at birth. Although more births will, as a proportion of the human species, be African, people in Africa will experience lower life expectancy than those in other regions.

In general, life expectancy has been increasing in giant steps across the world, driven by improvements in healthcare, education, quality of life and security, which at root owe their nurturing to the greater empowerment of women. But large inequalities remain, and this will be the case as much at the end of the present century as now. According to the analysis by Murray and colleagues, in 2100, life expectancy will still vary widely. Depending on where you live, it will range from 69.4 years to 88.9 years. Seven out of the ten countries forecast to have life expectancies less than seventy-five years will be in sub-Saharan Africa. The gap, however, is forecast to narrow.

Life expectancy aside, another consequence of the downturn in fertility is that the world population is greying. The average human in 2017 was 36.2 years old. By 2100 they will be 46.2, a whole decade older. Today, there are 1.7 billion over-65s. In 2100, this will increase to 2.37 billion. The figure for octogenarians is even more pronounced, increasing from 141 million to 866 million. Over the same period, the number of children under five will decline from 681 million to 401 million.

Comparing the number of births with that of people celebrating their eightieth birthday is salutary. In 1950, there were twenty-five births for each person turning eighty. In 2017 it was down to seven. In 2100 the figure will reach parity – just one birth for each person putting eighty candles on their birthday cake.

The ups and downs (and mostly the downs) of the world population make eye-watering reading, and that's before one starts to appreciate the economic consequences. The ageing population, combined with the decline in birth rates, means that there will be fewer people in any country, as a proportion of the population as a whole, who will be working,[20] generating wealth and paying taxes to support public services such as the care of the very young and pension provision for the elderly. In 2100, the countries with the largest pool of potential workers are forecast to be, in descending order, India, Nigeria, China and the United States. The number of workers in India and China, however, will be declining sharply. There is a solid relationship between working-age population and gross domestic product (GDP). Murray and colleagues use this to show that China will displace the United States as the world's largest economy by 2050, but they will change places again by 2100 as the working-age population of China falls. India, buoyed up by a vast (though shrinking) workforce, was the seventh largest economy in 2017 and will rise to third place in 2100, behind the United States and China and above Japan. Once again, the burgeoning population of Nigeria will see it rise from twenty-eighth spot in the GDP hit parade in 2017 to twenty-first in 2030, seventeenth in 2050 and ninth in 2100.

One of the main factors keeping GDP high is immigration from elsewhere. Immigration will see Australia, for example, steadily ascend the GDP charts from twelfth today to eighth in 2100, just one spot above Nigeria. The surging population of Israel, both through its high TFR and strong immigration, will see the country rise from thirty-fifth spot in 2050 to breach the top twenty – reaching sixteenth in 2100. Immigration is a quick and relatively easy way to sustain GDP, so governments of countries with falling birth rates might seek to boost GDP by liberalizing immigration policies.

Japan, though, will buck the trend. Despite massive shrinkage in population (more than halving from 128.36 million in 2017 to 59.72 million in 2100) and a reluctance to embrace immigration, its GDP rank will slip only from third in 2017 to fourth in 2100. It will maintain its position, in part, by increased participation of older people in the labour force. For the fortunate, increased life expectancy goes hand in hand with greater health, well-being and ability to work for longer. Other countries, such as the United Kingdom – and, more controversially, France – have sought to raise the age at which people will be entitled to claim a state pension.

Some countries have looked for ways to reverse the slump in TFR by introducing positive incentives to have children, such as generous parental leave and provision of childcare, but the effects are usually modest. In Sweden, for example, where the government actively pursued such policies, TFR increased from 1.5 in the late 1990s to 1.9 in 2019 – an increase, but still below replacement level. Other countries have pursued unwelcome measures such as restricting access to abortion. Such pro-natalist policies, however, are not effective because, even if they are sustained, they take at least a generation to filter through to what people regard as social norms. For example, after the reversal of China's two-decades-long 'one-child' policy, many young couples – each one of whom is an only child themselves – are allowed to have two children, but often choose not to. Furthermore, falls in TFR tend to dictate social norms for the future. Once TFR starts to fall, it is challenging to reverse that trend.[21]

In any case, it takes fifteen or twenty years for a new baby to become a taxpaying member of a country's workforce. Immigration is quicker. Countries that are slow to recognize the benefits of immigration might find that they have missed the bus. In less than a human lifetime, when almost all countries will be experiencing a drop in TFR and a downturn in population, governments of countries that formerly exported surplus people might wish for them to stay at home. As Murray and colleagues say, one consequence of a falling population is a decrease in the pool of innovators to drive economies. It takes a village to raise a child, but a civilization of hundreds of millions, even billions, to produce a Darwin or a Pasteur, a Gates or a Jobs, a Bezos or a Musk, a Newton or an Einstein.

The question that Murray and colleagues do not address (or do so indirectly) – and the question that is most pertinent to my present purposes – is why *now*? Given that the human population had been increasing exponentially for the past 10,000 years with no prospect, at the time, of change, one can perhaps forgive Paul Ehrlich his pessimism. But things have changed. One is entitled to ask why, uniquely in the entirety of human history and prehistory, is the human population on the cusp of steep decline?

The main acknowledged driver of the reduction in TFR in Murray and colleagues' detailed analysis is the reproductive and educational empowerment of women. Increasing take-up of secondary education of girls, and increasing supply of contraception to young women, is a cornerstone of their forecasting. The complete emancipation of women has still a long way to go, and in some places is occasionally thrown alarmingly into reverse, but it is a fact worth considering that female emancipation of any kind has only been recognized as natural and normal, and, even then, not everywhere, in the past century or so, for barely an eyeblink of human history. One should caution that such a recent development might just as easily be removed – witness the suppression of women's rights in countries such as Afghanistan, and the overturning of the constitutional right to abortion in the US. That a circumstance that seems so right and good and normal to some is anathema to others. On the other hand, when cast against the pageant of history, the increased empowerment of women seems to be happening all at once, in every country, rich or poor, regardless of its ethnic, religious or social background, or its system of government. The reduction in TFR is something that has happened despite the whims of authority. It has been sustained for the past half century and has been driven from the ground up. What is driving it?

The first factor to consider here is personal choice. Many couples want to have children, but given the choice, many will delay starting a family until they feel they are able to support one. This choice is primarily dictated by economics, and given the global decline in TFR, one has to look at the global economic outlook. The SARS-CoV-19 pandemic of the early 2020s dampened global trade, as did the financial crisis of 2008, but there are signs that these crises were set against underlying gloom. The global economy has been shaky since the turn of the millennium, and several factors may be at work. The Institute of Chartered Accountants in England and Wales reports[22] that in recent decades global trade has been boosted by the substantial movement of component parts for finished goods, some of which might cross borders multiple times before the finished item is ready for sale. Each stage is governed by a 'just-in-time' policy – that is, goods are made precisely when there is demand for them, rather than being made beforehand and stockpiled. This means that global trade is extraordinarily sensitive to shocks such as the SARS-CoV-19 pandemic and the financial crisis. A recent case has been the war in Ukraine, an essentially local conflict that has nonetheless sent shock waves through global energy markets and raised fears about global food security. Compounded with increased protectionism – itself driven by economic fears – such factors leave the global economy vulnerable.

A forecast issued in 2022 by the International Monetary Fund is grim,[23] to say the least, predicting a downturn in a third of the world's countries and a one-in-four chance that global growth could fall below a historic low of 2 per cent, perhaps as low as 1.1 per cent. The World Bank has warned of a global economic downturn cast against a picture of depression and insecurity. Slow growth (and the prospect of recession) affects 95 per cent of advanced economies and almost 70 per cent of emerging markets. The consequences of this will be increased poverty in some areas.[24] Sub-Saharan Africa – the source of much of the world's current population growth – is especially vulnerable. The consequences of increasing poverty, especially in sub-Saharan Africa, are hard to judge. Will women be able to exercise choice and restrict family size – or will the contraction in economies throw female emancipation into reverse, increasing family sizes?

I suspect that current economic uncertainty reflects a much deeper, long-term trend, and that's the decreasing accessibility of resources. *Homo sapiens* now dominates the Earth's natural economy, and has extracted so much from it that all the low-hanging fruit have been picked, and new resources are scarce. This might itself manifest, eventually, as a global drop in fertility. It's worth noting that at the beginning of the twentieth century, even households of modest means (in Britain, at any rate) could expect to have servants. The 'baby boomer' generation that grew up in the 1960s, as world population growth reached its peak, had access to far more resources, but were probably less likely to afford or employ staff. The generation that came after found it hard to afford their own home, irrespective of whether their establishment would come with a maid or a gardener. The World Economic Forum reports that millennials (those born between 1980 and 1994) will be the first generation, perhaps for many decades, who can expect to earn less than their parents.[25] As I write, housing in some of the world's major cities is at such a premium that young people, even if earning good salaries, cannot afford to rent anywhere to live, let alone buy a property of their own.[26] As the population peaks, there are fewer resources to go round. This in turn raises questions about whether reporting economic health in terms of increasing GDP – which equates economic health with consumption – is a desirable or even appropriate measure.[27]

It takes two to tango, and there is more to declining fertility than the reproductive choices of women. A less well-reported factor is a sharp decline in the reproductive health of men over the past few decades. For reasons nobody can quite fathom, human sperm count has fallen, both markedly and recently. Clinicians first started noticing this more than half a century ago, but the causes remain as mysterious now as ever.

Thirty years ago, Elisabeth Carlsen from Copenhagen University Hospital and colleagues[28] surveyed data going back fifty years to show that mean sperm count between 1940 and 1990 had not just dipped, or declined slightly, but *halved*. A much more extensive study published in 2022 by Hagai Levine of the Hebrew University of Jerusalem and colleagues showed an appreciable decline in sperm concentration and total sperm count in otherwise healthy men worldwide between 1973 and 2018.[29] The researchers called for action to 'recognize the importance of male reproductive health for the *survival of the human (and other) species*' (my emphasis). Whatever the causes, then, clinicians acknowledge the decline in sperm number and quality as an existential threat.

This study was important because it plugged a hole in the researchers' earlier work[30] that concentrated on North America and Europe, with the addition of Australia and New Zealand, by including data from South and Central America, Asia and Africa. The research answered a criticism that the decline in sperm count might be a feature of 'Western' or what we used to call 'developed' or 'industrialized' countries rather than the human species more generally.

Even in China, currently the world's most populous country, sperm count is declining – as deduced from a study of more than 300,000 healthy Chinese men between 1981 and 2019,[31] from which the researchers posted a 'serious reproductive health warning'.

What is the reason for this precipitous, global decrease? Some have linked it to a decline in the human male hormone testosterone, as well as increases in the incidence of genital anomalies and testicular cancer. Niels Skakkebæk of Copenhagen University Hospital (who was a co-author of the 1992 Carlsen study) and colleagues[32] hypothesize that increasing exposure to pollutants derived from fossil fuels might be one reason. Another factor that could affect results is that people are seeking to start families when they are older. If older men have fewer sperm than younger ones, this could bias the results. However, the studies take age into account – sperm count is dropping in men irrespective of age. At the same time, referrals for assisted reproduction (*in vitro* fertilization, or IVF) have increased. Since the birth of Louise Brown, the celebrated 'first test-tube baby' in 1978, more than eight million babies have been conceived in this way. But that's a drop in the bucket: Even today, only one in a million babies is conceived other than in the old-fashioned way.

If sperm count is declining in men of all ages worldwide who otherwise report as healthy, the cause must be something less definable and easy to spot and something pervasive. Constant, low-level exposure to derivatives of fossil fuels could fall into this category. Some studies have pinpointed pollutants that have feminizing effects on wildlife. It's inconceivable (if you'll excuse the word) that such substances will not have some effects on humans, perhaps by damping the production of testosterone.

Climate change could be a factor. It's well known that sperm production requires temperatures lower than those found in the body cavity. This is why the testes drop during puberty, to hang loose, outside the cosy confines of the abdomen. Lifestyle could be another. In their study on China, Mo-Qi Lv of Xi'an Jiaotong University and colleagues found that the decline in sperm quality was less marked in southern China than in northern China, where teenagers in particular tend to have higher body mass indices. That is, they are fatter.

As discussed above, Africa is the continent thought likely to experience the greatest increase in population in the coming century, as it has the youngest population. Yet data on sperm quality in Africa are sparse. They do, however, exist, and present evidence for a severe drop in sperm quality and concentration. One study[33] pointed up a high body mass index as a possible cause for low sperm count. Another, looking at data mostly gathered from Nigeria (the most populous nation in Africa),[34] reported a decrease in sperm concentration of 72.6 per cent (almost three quarters) over the past fifty years, citing a litany of possible causes such as obesity, disease, tobacco and alcohol consumption, and exposure to pesticides banned in other countries. Worryingly, mean sperm concentration in Africa (just over 600 million sperm per ounce) is hovering close to the minimum value considered by the World Health Organization as the 'cut-off value' (450 million per ounce), below which serious infertility is the result. If these results are borne out, then the population explosion expected in Africa might be significantly dampened.

Another possible reason is the stress caused by living in close proximity to other humans. For most of its existence, *Homo sapiens* lived in small, scattered groups. It is only relatively recently that our species has congregated in villages, towns and now huge cities. Some 55 per cent of the world's population is now urbanized, a proportion that is projected to increase to 68 per cent – more than two thirds – by 2050.[35] This trend is only likely to go one way as droughts, floods, crop failure and other manifestations of climate change drive rural populations to the closer confines afforded by urban life. For people to live on top of one another (literally so, in apartment blocks) and to pursue their daily activities so close to other people – such as in the Delhi street scene that so appalled Paul Ehrlich – is a relatively recent phenomenon. Living so close together is not our natural state. This might lead to a generalized increase in stress and could, according to at least one study,[36] lead to a decline in sperm count.

Overpopulation leads to environmental degradation, which leads to lack of resources, and all these could be weakening the global economy, driving down living standards, the desire for families among women and the sperm count of men who wish to become fathers. And behind all these things lurks the proverbial elephant in the room – that is, climate change. That's something I'll address in the next chapter.

# 8

# OVER THE EDGE

*Some ingenious writers have suspected that Europe was much colder formerly than it is at present; and the most ancient descriptions of the climate of Germany tend exceedingly to confirm their theory.*

EDWARD GIBBON,
*The Decline and Fall of the Roman Empire*

Climate change begins at home.

I live in the picturesque seaside town of Cromer in Norfolk, England. Cromer wasn't always on the coast. It's not mentioned as a settlement in the Domesday Book of 1086. In its place was a fishing village called Shipden – or, to give it its full name, Shipden-*juxta*-Crowmere, that is, Shipden next to the mere (or lake) of crows. The name of that possibly conjectural inland topographical feature gave its name to the modern town of Cromer and can still be found on the town's coat of arms, which depicts three crows on either side of a body of water. But what happened to Shipden? The answer to that can be given in a single word – erosion. Shipden has long been lost to the sea. Not so lost, however, that the underwater remains of the village did not pose a hazard to shipping. In 1888, Shipden's submerged church spire had to be dynamited after a pleasure boat from nearby Great Yarmouth ran aground on it at low tide. Shipden is not the only village lost to the sea in this area.[1] As the climate warms, as the level of the sea rises and as extreme events such as storms become more frequent and acute, this erosion is worsening.

We have already visited Happisburgh, a village a few miles along the Norfolk coast to the east of Cromer. This is where early relatives of *Homo sapiens* – possibly *Homo antecessor* – made a trail of footprints and left behind a set of exquisitely worked stone tools. In those days, though, these early tourists probably left their litter well inland, on the sandy bank of an ancient tributary of the river Thames rather than on the beach. Ever since, the approach of the sea has been inexorable. A village called Whimpwell once lay between Happisburgh and the sea. Whimpwell, like Shipden, has now completely disappeared beneath the encroaching North Sea. Between 1600 and 1850, the sea advanced some 275 yards, an average of three feet a year. The sea continues to take bites out of Happisburgh to this day, washing away farmland, livelihood, houses and homes.[2] Over the past twenty years, thirty-four homes in Happisburgh have disappeared.[3] The homeowners have had to move further inland – refugees from climate change, just a few miles from my front door.

Happisburgh is far from being an isolated case. Even further east along the coast, residents were alarmed in 2008 by a report – really, no more than a contingency plan – to stage a managed retreat before the sea, allowing six villages, hundreds of homes and twenty-five square miles of land to be flooded. The plan appears to have been shelved. But the problem has not gone away.[4]

The erosion eating away at the coast of Norfolk is one manifestation of a changing world. It's been going on since the end of the last ice age – a consequence of low cliffs made of soft sand and clay rather than hard rock, and the fact that the entire land mass of Britain is tipping over. In the ice age, Scotland was weighed down by a massive ice sheet. When this melted, the ground relaxed, leading to a slow uplift beneath the north-west of Britain, where the ice had once been sitting, a phenomenon called isostatic rebound.[5] The consequence is that the island of Britain is tipping, so that as the north-west rises, the south-east corner is slowly sinking.

Such events can be anticipated, planned for. What's worrying is that the erosion is speeding up. And extreme weather events are happening more often.

Perhaps the most extreme to hit Norfolk in living memory happened on the night of 31 January 1953, when gale-force winds and high tides inflicted the worst peacetime disaster ever to hit this part of the world.[6] A fierce northerly gale barrelled down the southern North Sea, pushing up a tide that was already very high – eight feet higher than predicted in Norfolk and fourteen feet higher than expected across the narrow sea, in the Netherlands. A few miles west of Cromer, in the neighbouring town of Sheringham, the storm surge overtopped seafront hotels, and homeowners were terrified as storm water poured down their chimneys. More than a hundred feet of the seafront was ripped away. In the village of Mundesley – between Cromer and Happisburgh – residents spoke of waves seventy-nine feet tall that swept entire buildings out to sea. To the east, an evening train leaving the seaside resort of Hunstanton for the nearby port town of King's Lynn was halted as the inrushing sea flooded the railway line. Moments later, the train itself was hit by an entire bungalow, torn from its foundations and washed along with the storm water. Not that the situation in King's Lynn was any better: Nine people drowned, and more than 3,000 homes were flooded there that night. The death toll in Norfolk was a hundred; it was more than 300 in eastern England. The toll in the Netherlands was more than 1,800.

Thankfully, nothing like this disaster has happened since – and if it did, modern weather forecasting might have given sufficient warning to allow people to flee for safety. There is no sense that the disaster of 1953 was the direct result of modern climate change attributable to the activities of humans. But with the changing climate making the weather less predictable, the ghosts of 1953 might yet come back to haunt this quiet, rural corner of England.

A 2018 report from the UK's Climate Change Committee noted that more than half a million properties, including 370,000 homes, are in areas at risk of damage by coastal flooding. Already, the cost of this damage exceeds £260 million each year. And it's going to get worse – by the 2080s, some 1.2 million homes will be exposed to flood risk, with 100,000 of these at risk of damage due to coastal erosion. And it's not just homes. By 2100, coastal flooding or erosion could have washed away 1,000 miles of road, 400 miles of railway lines, ninety-two railway stations and fifty-five old refuse landfill sites – leading to significant disruption to infrastructure and pollution of the environment. The report notes that the public is not well informed about the risks. A big issue is that coastal management is covered by a patchwork of organizations with differing responsibilities, so nobody has a view of the big picture. Nor are many people aware that protecting England's vulnerable coasts from climate-change-induced erosion and flooding will cost up to £30 billion.[7]

This story should give pause were one inclined to think that climate change is something that happens to other people, far away from England's tranquil but increasingly unquiet shores – people living, for example, in the low-lying Ganges Delta of Bengal, or on remote coral atolls in the Pacific Ocean. As the events of 1953 showed, nothing could be further from the truth. Climate change is happening everywhere, and it is happening now. Because of climate change, people all over the world will be on the move. You, too, might soon be among them.[8]

It's not just quaint seaside towns that will be affected by increased flooding as the sea level rises and extreme climate events become more common. Up to 200 million people (3 per cent of all humans) live in coastal cities that will be below mean high tide by 2100. Even if global temperature rise is limited to 3.6 °F above pre-industrial levels, this still translates into a sea-level rise of 15.4 feet in the long run and threatens land now occupied by 10 per cent of the global population – some 800 million people. If the temperature rise gets as far as 7.2 °F above pre-industrial levels, a billion people could be flooded out.[9]

Cities such as New York could be 6.5 feet underwater by the end of the century. Poignantly, the UN building itself could be awash. Although the upper levels would be useable, the UN would be an island surrounded by a moat, cut off from the rest of the city except by boat (the subways will be underwater) and even difficult to reach by air, as both LaGuardia and John F. Kennedy airports will be submerged. New York has already experienced a taste of the future. On 29 October 2012, when Hurricane Sandy hit the city, a storm surge of 8.8 feet coincided with high tides to inundate several subway lines, causing US$5 billion of damage.[10] More than thirty tropical storms and cyclones have hit New York since 1970,[11] and they are only likely to increase in frequency and ferocity.

New York is the most densely populated city in the United States. Having been built on a series of peninsulas and islands in a floodplain, it is exceptionally vulnerable to disasters like this. More vulnerable still is New Orleans, devastated by Hurricane Katrina at the end of August 2005. Much of New Orleans was already below sea level, protected by a series of dikes or levees, many of which were breached by the storm. The hurricane submerged 80 per cent of the city. As a result of the storm, 1.5 million people in the states of Alabama, Mississippi and Louisiana had to leave their homes. Around 40 per cent of them, mostly from Louisiana, were unable to return and had to resettle elsewhere, in some cases more than 800 miles away. The exodus from New Orleans was the largest single displacement of people in the United States since the Dust Bowl of the 1930s. It adds to an increasing catalogue of people internally displaced within the United States as a result of extreme climate events.[12]

The global picture is stark. Risk assessment consultancy Maplecroft estimates that 414 cities across the world (with a combined population of 1.4 billion) are at high to extreme risk from pollution, lack of water supplies and climate-change-related disasters.[13] They include Los Angeles, Mexico City and Lima. European cities include Naples, Athens and Rome. Ninety-nine of the top hundred are in Asia, and include Wuhan, Tokyo, Manila, Lahore, Karachi and Delhi.

Top spot in this chart of imminent catastrophe is Jakarta, the capital city of the populous island nation of Indonesia, which, in addition to congestion, pollution, overzealous extraction of groundwater and the other problems that plague world cities, is also subsiding as a result of climate-change-induced sea-level rise and seismic activity. The northern parts of Jakarta, on the Javanese coast, have already been abandoned to the sea, and a third of the city could be submerged by 2050. There are proposals to move the seat of government to higher ground on the island of Borneo.[14]

Although cities in Africa are less exposed than others to natural hazards, cities such as Lagos, Kinshasa and Nairobi are uniquely vulnerable to climate change by virtue of relatively poor infrastructure. Here, heat will be the killer. At some point in the relatively near future, some of these tropical megacities will become unliveable.

Sometimes it's nice to have a bit of warmth.

In the summer of 2003, my family and I spent an idyllic holiday by the sea in Wales, where it was warm, sunny and dry for two whole weeks. For anyone who knows Wales, such an uninterrupted period of warmth and dryness is remarkable. The climate of Wales, on the western edge of the island of Britain, is proverbially damp and mild, even in summer. What we did not know at the time was that the heatwave of 2003 was the most pronounced seen anywhere in Europe for 500 years. The UK Meteorological Office reported that it killed 20,000 people in Europe, 15,000 of them in France, which ran out of space to store dead bodies in its mortuaries[15] – adding that summers as hot as that of 2003 could happen every other year by 2050.

Summers have been getting hotter. Britons of a certain age (such as me) recall the amazingly hot summer of 1976, green in the memory because it was so extraordinary. But 1976-style summers have been getting more frequent, with hot summers in 1995, 1997, 2003, 2006 and 2022. Heatwaves are more pronounced in cities due to the 'heat island' effect, in which buildings and roads reflect and radiate more of the Sun's heat than fields and forests. This can be deadly. Residents of Chicago might recall July 1995, when a heatwave claimed around 700 lives in the space of five days.[16]

And still the world continues to heat up. On 7 May 2023, almost twenty years after our family beach holiday, the Southeast Asian nation of Vietnam recorded its highest ever temperature of just over 111.2 °F. The thermometer in Thailand, meanwhile, peaked at 112.3 °F, whereas a town in eastern Myanmar reported a temperature of 110.8 °F, the highest temperature for a decade.[17] It's normal for the weather in this region to become very hot just before the Southeast Asian monsoon, but nothing like this intensity had been experienced before. India and Bangladesh also reported unusually high temperatures in April and May 2023. As hundreds of millions of people queued to cast their vote in India's general election in May and June 2024, dozens of people died (including election officials) as the temperature climbed to 121.8 °F, the highest temperature ever recorded in that country.[18]

Europe is not exempt. Spain reported its highest ever temperature for April in 2023, with a peak of 101.8 °F at Córdoba, in the south of the country.[19] Spring faded before a summer during which the Mediterranean basin sweltered under ferocious heat; in concert with a prolonged dry spell, that heatwave led to widespread wildfires that destroyed homes as well as the tourist industry on which many places in the region depend. The calendar year of 2023 was the hottest since records began in 1850.[20] The temperature continues to rise, and records continue to fall, so that by the time you read this, even these extreme temperatures might well have been exceeded.

Excess heat can be lethal, especially to the very old and the very young, who are less able than most to regulate their body temperature.

A potent addition to the mix is humidity. Humans, having evolved in the hot, dry savannahs of Africa, have a natural way of cooling – sweat. The evaporation of the water in sweat from the skin requires energy to turn it into vapour. The energy to do this is drawn from the skin, so the effect of evaporation is to cool the body. But this only works if the air above the skin is dry and isn't already saturated with water vapour. In 100 per cent humidity, the water cannot evaporate, because it has nowhere to go, and the cooling effect cannot happen. The combined effect of heat and humidity is measured by the so-called wet-bulb temperature, which is the temperature experienced at 100 per cent humidity, when cooling by evaporation is no longer possible. Sweating means that humans, if sufficiently hydrated, can tolerate temperatures in dry heat in excess of 104 °F for a period. But exposure to more than 95 °F in 100 per cent humidity for more than six hours is likely to be deadly even for fit, healthy humans.[21] In the current climate, such wet-bulb temperatures rarely top 87.8 °F, anywhere on Earth. But this could change.

The Persian or Arabian Gulf – often known just as the Gulf – is a shallow, nearly landlocked arm of the Indian Ocean that separates the Arabian Peninsula from south-west Asia. The nations surrounding it are the source of much of the world's oil. Gulf cities such as Doha in Qatar and Dubai in the United Arab Emirates are bustling international business and tourism hubs. This is already a notoriously hot part of the world, but the proximity of the shallow, quickly warmed, fast-evaporating waters of the Gulf makes it humid, too. One scenario,[22] projecting heat and humidity in the Gulf for the final thirty years of the century (2071–2100), suggests that wet-bulb temperatures could exceed 95 °F in Dubai, Doha, Abu Dhabi and other Gulf locations several times. What counts as a scorchingly hot and humid summer in the Gulf now would, by then, be quite normal. In other words, the Gulf could become unliveable for humans without extraordinary mitigation, such as building cities underground. Gulf cities are tolerable now for those fortunate enough to have air conditioning. Matters are different for those in poverty or constrained to spend time outdoors.

In Mecca, Saudi Arabia, towards the end of this century, temperatures could top 131 °F with a wet-bulb temperature of 89.6 °F. Overheating is already a problem for pilgrims on the Hajj, obliged to pray outdoors during daylight hours near Mecca. It is perhaps inevitable that the fervour, excitement and crowding of the Hajj have historically led to casualties, but increasing heat in an already hot country has made matters worse in recent years. In 2024, hundreds of pilgrims died in heat exceeding 122 °F: The temperature in the Grand Mosque hit 125.2 °F.[23]

In relatively impoverished Yemen, wet-bulb temperatures around Aden could hit 91.4 °F, putting elderly people and children at risk of death. According to Elfatih Eltahir and Jeremy Pal of the Massachusetts Institute of Technology, writing in the journal *Nature Climate Change* in 2016, 'A plausible analogy for future climate for many locations in Southwest Asia is the current climate of the desert of Northern Afar on the African side of the Red Sea [opposite from Yemen], a region with no permanent human settlements owing to its extreme climate.'[24] Extreme heat could depopulate large areas of the Middle East, where the current inhabitants will have to move or die.

The Gulf is not an isolated case. At first glance, there is no greater contrast with the Gulf than the plains of northern China, one of the most densely populated regions of the world, and home to 400 million people. Whereas the Gulf is surrounded by hot, dry deserts on all sides, northern China is well-watered and richly irrigated. Expansion of irrigation in the region moistens the air above the surface – and that's the problem. One projection of climate change in northern China puts wet-bulb temperatures higher than would be tolerable for people working outdoors, such as agricultural labourers.[25] It's a similar story for India, Pakistan and Bangladesh, home to about a fifth of the human population, many of whom work outdoors farming the fertile Indus and

Ganges valleys. Southern Asia is already experiencing heatwaves of extreme intensity. Warm, moisture-laden air off the Bay of Bengal and the Arabian Sea combines with moist air from intensive, irrigated agriculture and low-lying land to make a muggy brew of saturated air. By the end of this century, the plains of northern India will be regularly subjected to wet-bulb temperatures exceeding 95 °F, beyond which human life is essentially unsustainable.[26]

Globally, almost a third of the population is exposed to deadly heatwaves for more than twenty days a year. By 2100 this could rise to almost half, even with rigorous control of greenhouse-gas emissions, but rising to three quarters if emissions continue unabated.[27] In addition to the Gulf and parts of India and China, the south-east of the United States and coastal West Africa are in the frame for substantial increases in exposure to potentially lethal wet-bulb temperatures.[28]

West Africa is a particular cause for concern. The increase in exposure to moist heat over the next few decades in coastal regions of countries such as Nigeria will coincide with a ballooning population with limited means of mitigation or protection. All in a part of the world experiencing environmental degradation, the encroachment of deserts away from the coasts and relatively fragile governments that may be in conflict with armed insurgents. Heat, too, may generate conflict – people get angry when they are hot. The link between climate change and the incidence of armed conflict is highly debated. However, a recent study suggests that the influence of climate change on conflict is 'both substantial and highly statistically significant'.[29]

Climate change, rapid population growth and poor prospects at home due to economic and political uncertainty are all factors that drive migration. In her book *Nomad Century*, Gaia Vince[30] suggests that over the coming decades, the migration of people from the torrid Global South to more equable climates in the north will be inevitable, and, with the right mindset in the north, welcome, particularly as many countries in the Global North will be experiencing rapid contractions in their native populations.

Migration is, of course, the natural state of humanity. Before the invention of agriculture and the beginnings of sedentary life, which accounts for only about 3 per cent of the entire time our species has been on the planet, bands of *Homo sapiens* never settled anywhere for long, always moving on to where the grass was greener. The same was true for its ancestors.

Although members of the human family have always been restless, there are two great episodes of migration that stand out. The first was about two million years ago, when *Homo erectus* left Africa for the first time, expanding into Eurasia and diversifying into many different species, from Neanderthal cave-dwellers in Europe to the hobbits of south-east Asia, *Homo antecessor*, and *Homo heidelbergensis*.

The second migration occurred – probably in several waves – between about 125,000 and 50,000 years ago, when *Homo sapiens*, a species that had hitherto been confined to Africa, spread into Eurasia, eventually replacing all the other members of the human family.

We are on the cusp of witnessing a third human migration from Africa and into Eurasia. These refugees from an increasingly hostile climate will make their way north. No legislation, no patrols of boats in the Mediterranean or the North Sea, will be able to stop them.

Not everyone will choose to migrate, however. People living in the increasingly hot, humid and occasionally flooded parts of the world will, in general, try their best to stay where they are. If they do, billions of them may perish, accentuating the eventual worldwide decline in the human population that will occur around the end of the century. Others, though, will adapt, seeking to build cities increasingly cut off from the hostile environment outside. The flood tide of migrants reaching the north, as numerous as they might seem, will be in the minority.

In the past two chapters I've sought to enumerate the challenges now facing the human species. In Chapter 7 I looked at the slowdown of human population growth. The human population will peak in the current century and then start to decline. Some predictions suggest that by 2100 there will be fewer humans than there are now. It's worth reflecting on this for a moment. Ever since the dawn of agriculture 10,000 years ago, the human population has only ever grown, not shrunk. As far as anyone knows, there have been no significant reverses, except possibly as a result of worldwide pandemics such as the Justinian Plague of the sixth century and the Black Death of the fourteenth.

The current reversal cannot be put down to any single identifiable cause. Or, rather, there are several causes, but the relationships between them are, so far, uncertain. On the positive side, the emancipation of women in many parts of the world over the past century or so has contributed to increased educational attainment, health, well-being and longevity. This is related to progress in science and medicine made over the past two or three centuries, and what demographers call the 'demographic transition', from families having many children, many of which were expected to die in infancy from diseases that are now easy to prevent, to them having fewer children, most of which, thanks to better hygiene, healthcare and innovations such as vaccination, can be expected to reach adulthood.

The decline in the number of children per family has led to a situation, in many countries, of people having hardly any children at all, resulting in an ageing population. The world population is still increasing, and will do so for several decades, leading to its own pressures, but is likely to shrink as the tumultuous, crowded twentieth and twenty-first centuries give way to the twenty-second. The consequences of a population that is increasing in the meantime include climate change and habitat degradation, as well as a reaction of the population itself to that overcrowding (manifested in increased migration), and unexplained phenomena such as the worldwide reduction in human sperm count.

In some ways, increasing populations are a good thing. They drive economies and innovation. Scientific and industrial productivity depend on a large and increasing supply of human brain power. The current general measure of economic health, GDP, is predicated on the assumption of growth. But one can have too much of a good thing.

In my view, the cause of the coming decline in human population is that the resources required to maintain economic growth are becoming rare and more expensive to find, process and distribute without consequences for general health and the environment that might outweigh the benefits. For the past few decades, the ups and downs of the global economy have been against a picture of general stagnation: GDP might therefore no longer be a suitable way to measure economic health, given its assumption of ever-greater wealth and prosperity. One study suggests that the Earth can sustain seven billion people at subsistence levels – but this total has already been exceeded, and current demands for a good quality of life for all would stretch the capabilities of our resource base between two- and sixfold.[31] It seems clear to me that all the trends discussed in Chapter 7 – from the stagnation of the global economy to the reduced aspirations of millennials compared with their parents, to the decline in sperm count – can, at root, be attributed to this single factor.

What I did not discuss in Chapter 7, but addressed in the current chapter, is the threat to the well-being of humans posed by climate change. This is rapid and severe, and whether by heat, by drought or by flooding, or whether by war precipitated by these disasters, climate change poses an existential threat to *Homo sapiens*. None of the projections that portray the coming decline in the human population explicitly take the effects of climate change into account.

It is possible that the declining fertility rate of humans reflects some inherent acknowledgement by them of the threats posed by climate change. Given current evidence, it is hard to say whether this acknowledgement is manifested as conscious decisions, dictated by economic circumstances or career aspirations in which people decide to defer having children, or physiological ones, such as the decline in sperm count as a reaction to overcrowding, stress or some other factor. Very likely it is a mixture of both.

Of course, much of this discussion is based on projections – both of human population trends and of changes in climate. We already know that prediction is an inexact science. So, it is possible – just possible – that none of the doomsday scenarios I've discussed here will come to pass. However, as time goes on, and the predictive models are refined with better and more comprehensive data, the picture seems to only get worse, not better.

When Paul Ehrlich wrote *The Population Bomb*, the world population was a fraction of the current total. Innovations such as the Green Revolution in agriculture (which I shall discuss later) and trends (such as increased female emancipation and changing norms about family size) that could not have been seen clearly at the time allowed the population to grow without the famine and disasters that Ehrlich predicted. In a sense, *Homo sapiens* has merely postponed the day of reckoning: The world now contains more people than it can support sustainably for any length of time, and the population is still growing.

However, in a few decades it will level off and start to shrink, possibly very rapidly – something that was not predictable in the 1960s, when the world population reached and passed its highest rate of growth. The decline in human population would be dramatic even in the absence of climate change. Once that is added to the mix, the contraction in the human population is likely to be a good deal less comfortable than it might be or needs to be.

# 9

# FREE FALL, AND AFTER

*On the slightest touch the unsupported fabric of their pride and power fell to the ground. The expiring senate displayed a sudden lustre, blazed for a moment, and was extinguished for ever.*

EDWARD GIBBON,
*The Decline and Fall of the Roman Empire*

Predicting what the world's population will be like by the end of this century is hard enough, even without taking the additional hazards of climate change into account. After all, nearly all the people who will be alive to face the challenges of the twenty-second century are yet to be born. Perhaps the most we can say is that the world will house, in varying degrees of affluence or squalor, approximately the same number of people as it does now, but a greater proportion of them will be African, or of African descent.

Looking even further into the future is even riskier. One such study[1] dares to extrapolate current trends in fertility and life expectancy as far as 2300 and finds that the trend is sharply downwards. Even assuming a generous TFR of 1.5 to 1.75 would lead to a drop in world population to between 2.6 and 5.6 billion by 2200, and between 0.9 and

## FREE FALL, AND AFTER

3.2 billion by 2300. The last time the world's population was a billion was in 1804, during the Napoleonic Wars.[2]

Looking even further ahead than that would seem to be foolish indeed – but it can be done. An audacious thought experiment by a Princeton astrophysicist, J. Richard Gott, looked ahead to predict when our species might become extinct.[3] Gott doesn't extrapolate fertility or mortality or prognosticate about the effects of climate change. Instead, he makes an argument from probability.

Once upon a time, conventional wisdom had it that the Earth was the centre of the Universe, with the Sun and planets revolving around it. The Polish astronomer Nicolaus Copernicus (1473–1543) reasoned that the Earth moved around the Sun instead. This was the first step in removing any sense that the Earth occupied a privileged position in the Universe. As time went on, it became evident that the Sun wasn't at the centre of the Universe either, but an ordinary star among billions, in the quiet suburbs of an ordinary galaxy among trillions, with no particular claim on any privilege whatsoever. Over the past thirty years or so, it's been found that many stars – perhaps most – have systems of planets, so the Solar System is in no way unique.[4] In short, the Earth does not occupy a privileged place in the Universe, and, by extension, neither do we.

Gott's idea was to take the Copernican principle and apply it to time, rather than space. He reasoned that no intelligent observer (you, for instance) is likely to live at any particularly special moment of human history.

Gott started his argument small, recalling his own visit to the Berlin Wall[5] in 1969. Applying the Copernican principle, he reasoned that his visit was most unlikely to have fallen either near the Wall's construction, in 1961, or its demolition. He assumed that his visit occurred during the 95 per cent of its existence in the middle – neither in the 2.5 per cent of its existence near the beginning, nor in the 2.5 per cent near the end.[6] But when would the Wall be demolished? A simple calculation, based on his one random visit, put the Wall's destruction between 0.2 and 312 years after his visit – applying what statisticians call 95 per cent confidence limits. In the event, the Wall fell twenty years later, in 1989, well within Gott's admittedly broad limits.

Gott then applied his reasoning to the whole of human history. Imagine human history as a line, beginning with the earliest known appearance of *Homo sapiens* approximately 315,000 years ago, and its end at some time in the future. Plainly, you are living somewhere in between. But when? You have no idea, at least initially, whether you are living near the dawn of the human estate (in which case human history will extend much longer than we might project) or at some point near humanity's demise (in which case our tenure will be much shorter). You are somewhere in the middle.

As with the Berlin Wall example, let's assume that you are living during the 95 per cent of human history in the middle – you are more than 2.5 per cent of human history away from either the birth or the death of our species. To be close to one or the other of these epochal moments would confer a specialness on you that you neither want nor (in statistical terms) deserve. From this, it's easy to work out that the human species will become extinct at some point between 8,077 and 12,285,000 years in the future.[7]

## FREE FALL, AND AFTER

Leaving aside the undoubted fact that the achievement of *Homo sapiens* in all spheres, not least in population size, dwarfs that of any of its antecedents, human extinction more than twelve million years hence does seem generous. No other species in the human family lasted anything like as long.[8] For example, the first signs of Neanderthals – the closest relative of *Homo sapiens* – can be seen from fossils at the Sierra de Atapuerca in Spain dating back more than 300,000 years,[9] but the species went extinct soon after its encounter with *Homo sapiens*, around 40,000 years ago,[10] making a total residence on Earth of approximately a quarter of a million years. One could argue that Neanderthals might have lasted longer had they not had the bad luck to run into our own species.

*Homo erectus* – ultimately the antecedent of both species – lasted longer. It appeared around two million years ago and disappeared, very roughly, 100,000 years ago,[11] a period much longer than the Neanderthals but far short of a multimillion-year span. In general, mammal species (which include *Homo erectus*, Neanderthals and us) have a span of no more than one to two million years or so before either going extinct or, perhaps, evolving into something recognizably different.

What about the sharp end – the prediction that human extinction might be as soon as 8,000 years or so in the future? This is instructive, because it forces you to wonder whether you might have a privileged position in history. This is because you are living during the exact time, unique in the whole of the human career, when the rate of population growth is slowing and will go into reverse. From this I would argue that your position is located much closer to the end than the beginning of human history, and I'd venture – with suitable hand-waving – that *Homo sapiens* will disappear from the Earth within the next 10,000 years or so.

I do have yet another reason for pessimism, and it comes from an unlikely source. In 1994, *Nature* published a terse, highly technical paper from a small team of theoretical evolutionary biologists. It was just two pages long, and coined a concept called 'extinction debt'.[12]

An important cause of the extinction of any species is the destruction of the habitat it needs to survive. A species of bird that nests in tall trees will not be able to survive if all the trees are cut down and replaced with grassland. Habitats such as forests are not usually cut down all at once, but progressively dissected and subdivided by roads, and cut down in lots, leaving a network of smaller woodlands, separated from one another as if they were islands separated by ocean. What happens to the species left in these ever-smaller spinneys?

Species that are always rare are prone to extinction by random mischance. However, even species that are common in any given patch of woodland will, after a certain amount of habitat destruction, be marked for inevitable extinction, even if that crucial tree is felled when the species seems to be in good health and there seem to be plenty of other trees around. Such common species – those that stand and fight in any patch of habitat and appear to be in no danger – are called dominant competitors, and it turns out that dominance does not make them immune from inevitable extinction. Rather, the reverse is true. Even if such a species occupies 10 per cent of a pristine habitat, and 10 per cent of the same habitat is randomly chosen for destruction, that dominant species' occupancy of the remaining 90 per cent might seem, on the face of it, to assure its safety.

But this is not so.

'Surprisingly,' the researchers say, 'the destruction of random sites has the same eventual effect as selectively destroying *precisely* those sites occupied by the dominant competitor [my emphasis].'

Even if more generations were to pass since that one, crucial episode of habitat destruction, extinction will befall the species, paying off what the researchers call its 'extinction debt'. It might not know it, but it's a dead species walking. The paper closes with what reads, in retrospect, and considering everything I have written so far in this book, as a dire warning for our own species.

> Although it is well known that habitat destruction causes extinctions, our results warn that an unanticipated effect of habitat destruction may be the selective extinction of the best competitors. *These species are often the most efficient users of resources and major controllers of ecosystem functions* [my emphasis].

If it can be described as anything at all, *Homo sapiens* has become the dominant competitor in the single patch of habitat that it occupies – the Earth. So dominant, in fact, that it drove all the competition to extinction long ago. The degree to which humans have appropriated the Earth's resources is staggering. It has been estimated that between 25 and 40 per cent of all the products of photosynthesis by green plants on Earth have been sequestered by humans.[13] *Homo sapiens* and its domestic animals comprise 96 per cent by mass of all mammal species. All other mammals, from aardvark to zebra, must squeeze into the remaining 4 per cent. Out of every ten birds you might meet, seven (on average) are farmed poultry.[14] As I showed in Chapter 8, *Homo sapiens* is progressively rendering large sections of the planet uninhabitable not just for other species, but for itself. Nobody knows when or where that rainforest tree was felled whose destruction finally marked humanity's card. Perhaps it hasn't been cut down yet but remains entire and intact in the deeper recesses of Borneo or Amazonia. If so – woodcutter, spare that tree!

But how would we know which was that crucial tree? My guess is that it was cut down some time ago, and the fate of humanity has already been sealed.

It's not as if we haven't been warned.

Rapa Nui (Easter Island) in the eastern Pacific Ocean is one of the remotest habitable spots on Earth. The first European to set foot there was the Dutch explorer Jacob Roggeveen, in 1722. Rather than the lush, palm-fringed prospect one usually associates with islands in the South Pacific, the vista that greeted him was a treeless wasteland.

It had not always been that way.

By the time Roggeveen visited, the once-productive environment had been destroyed by the inhabitants.[15] Between the eleventh and seventeenth centuries, rival clans on the island strove to outdo one another by erecting the gigantic statues or *moai* for which the island is now famous.

In so doing, they drove all the native land birds on the island (there were at least six species) to extinction. Twenty-five species of seabird once bred on Rapa Nui. Now there is only one.

And they cut down all the trees.

Today, only forty-eight native species of plant live on Rapa Nui, the largest a shrub barely seven feet tall. Twenty-two other species have vanished, including what was once the tallest species of palm in the world, outdoing the largest extant species, the Chilean wine palm, which can grow up to sixty-five feet tall and have a girth of three feet. All were chopped down, to the very last. Once the trees and birds had all gone, the islanders were without the wood they needed to survive – to make the ocean-going canoes they used to hunt dolphin; without sources of fibre, construction materials, edible fruits or firewood. The consequence was mass starvation, a sharp decrease in population and a descent into civil war and cannibalism. Unable to find trees sturdy enough to build shelters, many islanders took to living in caves, partly sealing up the entrances as a defence against hungry neighbours. Writing in 1774, Captain James Cook described the remaining islanders as 'small, lean, timid and miserable', just scratching a living as subsistence farmers amid the megalithic hubris of their ancestors.[16]

One is entitled to ask whether the person who felled the last tree on Rapa Nui was conscious of the finality of the act. 'I have often asked myself,' wrote Jared Diamond,[17]

> what did the Easter Islander who cut down the last palm tree say while he was doing it? Like modern loggers, did he shout 'Jobs, not trees!'? Or: 'Technology

will solve our problems, never fear, we'll find a substitute for wood'? Or: 'We need more research, your proposed ban on logging is premature'?

Once the islanders had felled all trees large enough to build ocean-going canoes, they had effectively cut themselves off from the rest of the planet. They were alone, totally isolated and thrown on their own dwindling resources. The parallels between Rapa Nui and the Earth are perhaps too obvious to belabour.

Over the next few thousand years, the human population will sink to a level that is ultimately unsustainable, and extinction will beckon. I'll now return to the Karenina Principle outlined in the prologue: All happy, thriving, abundant species are the same, but each species facing extinction does so in its own way. Therefore, it's extremely difficult to be precise about how or where the very last human will meet its fate, even more so than estimating the time of the ultimate demise of humans. That having been said, it's possible to sketch out some general pointers from what we already know about the extinction of a species.

Leaving aside the occurrence of some worldwide calamity that wipes out millions or billions of people all at once (worldwide nuclear conflagration, an asteroid impact, robots on the rampage or a truly ferocious pandemic), extinction happens to groups of animals or plants that have already shrunk in number to become vulnerably small – a few hundred individuals at most. Very small populations face existential threat in a way that large ones never do, because there is safety in numbers. Small populations might be snuffed out by unseasonable weather, for example, or the sudden depletion of some vital resource. It's easy to imagine a species of, say, orchid, so rare that it's confined to a single patch of woodland, being driven out of existence in the course of a new construction project. It could have been that the orchid was once widespread, but a combination of factors has broken up the once-continuous population into an archipelago of tiny islands, or habitat patches, each remote from the others, and each subject to its own whims of fortune. First one patch of orchids disappears, then another, until there is only one left, and then – none at all.

Small populations face other threats, too, beyond the increased risk posed by random unfortunate events. When populations are small, there is an increased risk that otherwise unhealthy traits will come to the surface and become established in the community. As I discussed earlier, *Homo sapiens* is no stranger to the genetic problems associated with small population size. When unhealthy mutations become more common in a population, the effect is to decrease that population's fitness.

Here I should add a few words about what 'fitness' means, in genetic terms. In the language of evolution, 'fitness' means the capacity for individuals – or pairs of individuals – to produce offspring that will survive to adulthood and so be able to breed themselves. A population that is less fit will produce fewer healthy offspring. It stands to reason that when unhealthy mutations become common, fitness will decrease. The population will produce fewer offspring that will survive than they would otherwise – say, had the population been larger – and extinction becomes a risk.

Random threats and the decrease of fitness are not the only woes faced by small populations. When populations become small and fragmented, choosing a suitable mate becomes a challenge. It's always an advantage to select a partner from a different community from one's own. This will keep the worry of inbreeding at bay, reducing the risk that unhealthy mutations will surface and decrease the population's fitness. But when populations become scattered, the survivors ever more distant from one another, the quest for mates becomes ever more difficult, and, eventually, impossible.

All this sounds rarefied and theoretical, but there is a model for how the future will play out for *Homo sapiens*. Suitably, and perhaps poignantly, it comes from humanity's closest known extinct relatives, the Neanderthals.

As a reminder, the Neanderthals first appeared a little before 300,000 years ago. A population of hominins with signs of the characteristic heavy-browed, long-brained features of Neanderthals lived in the Sierra de Atapuerca in northern Spain at around that date. Neanderthals lived in Eurasia, from Spain in the west to Siberia in the east, and from the Russian Arctic in the north to the Levant in the south until, at the latest, 40,000 years ago.[18] Modern humans – *Homo sapiens* – began to appear in numbers in Eurasia from about 45,000 years ago, and the two species overlapped in time for between 2,600 and 5,400 years, after which period Neanderthals vanished.

The presence of *Homo sapiens* was, ultimately, the cause of the disappearance of the Neanderthals, as it was for the extinction of every other hominin on the planet. Not because modern humans were in any way superior, but because there were more of them. Neanderthals were simply overwhelmed by the new arrivals.

Neanderthals have been known since 1856, when bones of a strange and hitherto unknown kind of human were discovered in a cave in Germany. Although the skull of what was later recognized as a Neanderthal had been found in Gibraltar in 1848, extinct varieties of human were then unknown to science, and debate raged about whether the remains were in fact a new species, or from some pitifully deformed soul from more recent times. There was even the suggestion that the remains were of a Cossack soldier who perished during the Napoleonic Wars. Since then, many Neanderthal remains have been found, and it's a wonder of modern science that it is now possible, in certain favourable circumstances, to extract the DNA – the genetic material – from their bones and teeth.[19] In recent years this has given us a picture of the genetics of Neanderthals for most of their history, allowing us to chart their population structure, and shed light on their lives and times.

It seems clear that Neanderthals were never common. Throughout their history, they had lower genetic diversity than any modern human group. Bear in mind that modern humans are remarkably homogeneous, genetically: That Neanderthals had even lower genetic diversity than us reflects both their small population sizes, as well as the possibility that they nearly became extinct several times in their history, re-establishing themselves from surviving founder populations after each near-miss brush with eternity. For example, genetic evidence indicates that many

Neanderthals living more recently than 100,000 years ago descended from only a few progenitors, suggesting a genetic 'bottleneck' in which most Neanderthal populations had become extinct – the founder effect in action.[20] Again, there is evidence that very late in their history, many Neanderthal populations in Europe disappeared, possibly as a result of the worsening cold of the Ice Age, and were replaced by other Neanderthals moving in from warmer refuges in southern Europe and western Asia.[21] As a result of this tenuous hold on existence, Neanderthals had 40 per cent lower fitness than modern humans, on average.[22] This means that they were not good at producing offspring, and the offspring they did produce were less likely to reach an age where they were capable of reproduction themselves. Their low population size threw up various congenital abnormalities, such as deformities in the vertebrae and the retention of milk teeth into adulthood, with far greater frequency than is seen in modern humans.[23]

Inbreeding was common. DNA from the toe bone of a Neanderthal female from Denisova Cave in the Altai Mountains of southern Siberia shows that her parents were as closely related as half-siblings, and mating between close relations was common in her immediate ancestry.[24] Another study, of no fewer than thirteen Neanderthals from the Altai, put a spotlight on their communities in remarkable detail. Of the remains of eleven individuals from a cave called Chagyrskaya, two were a father and daughter, and another two were close cousins.[25] All were highly inbred, and the total size of the community was probably around twenty individuals at any given time. However, there was also evidence that most females in the community came from other groups. In common with primates generally, males tend to remain with the communities they were born into, but females tend to migrate elsewhere.

The migration of females between communities was vital to prevent Neanderthals becoming even more inbred than they were. It also promoted fellowship between otherwise isolated and scattered family groups. It was this factor – female migration – that probably explains why Neanderthals lasted as long as they did. Ever dancing on the edge of oblivion, the movement of females between communities was just enough to keep the flame alive. When modern humans moved into Neanderthal territory, one effect was to separate individual Neanderthal communities from one another, inhibiting the flow of females between them.[26] Increasingly isolated, Neanderthals resorted to inbreeding – or interbreeding with modern humans. Either course led to extinction of the Neanderthals as a separate species, leaving a trace of their DNA in all modern humans living today who do not have an exclusively African ancestry.

The long, sad story of the Neanderthals paints a picture of what life might be like for the last members of *Homo sapiens* on Earth. A collapsing population, coping with chronic lack of resources and the additional ravages of climate change, will become divided and scattered into increasingly isolated fragments. First one tiny population will wink out, and then a second, until, eventually, the last human in the last remaining group of *Homo sapiens* will disappear.

This will happen within the next 10,000 years.

Unless . . .

# PART THREE
ESCAPE

## 10

# THE FUTURE IS GREEN AND FEMALE

*If we contrast the rapid progress of this mischievous discovery [gunpowder] with the slow and laborious advances of reason, science, and the arts of peace, a philosopher, according to his temper, will laugh or weep at the folly of mankind.*

EDWARD GIBBON,
*The Decline and Fall of the Roman Empire*

I make no apology for spending the past few chapters painting a gloomy picture of humanity's future. A combination of problems both intrinsic (low sperm count, genetic homogeneity, economic woes) and external (the threats that human activity pose to the Earth's habitability) are conspiring to shorten the prospective tenure of *Homo sapiens* on Earth. And that's even without threats that I haven't discussed at all, such as nuclear war, the impact of a large asteroid or comet, or the unintended consequences of technological innovations such as artificial intelligence.

On the other hand, none of the doomsday scenarios I've outlined so far have yet come to pass. Like the Ghost of Christmas Future in Charles Dickens's *A Christmas Carol*,

I am showing you a world that might happen, were we to continue doing what we are doing. There is no certainty.

In this, the final part of the book, I shall attempt to sketch out what one might call a Hollywood ending. There could be a way out, a way to prolong the existence of *Homo sapiens*, potentially to reach the more distant end of Richard Gott's prediction, millions of years into the future. It rests on human ingenuity: its capacity, even when it seems that it is boxed into a corner and no salvation is possible, to break free.

My proposed solution is the colonization of space, whether the surfaces of other bodies in the Solar System such as the Moon or Mars, or the interiors of modified asteroids, or completely artificial orbiting habitats.[1] Given what we know of human ingenuity, there could be other ways out, too, such as 'terraforming' other planets like Mars or Venus, so that they become habitable to humans without special breathing equipment or spacesuits; or even large-scale editing of the human genome to remove undesirable traits or adapt humans to conditions hitherto unachievable by natural means, such as being able to breathe underwater, or thrive indefinitely in conditions of zero gravity,[2] or survive unaided in other extreme conditions. However, for reasons I'll explain later, my solution must be achieved within the next two centuries for human extinction to be avoided. Some solutions lie so far in the future (terraforming) or are likely to remain constrained by ethical concerns (large-scale genetic engineering) that I shall not discuss them further.[3]

The drive to colonize space will begin with changes *Homo sapiens* make here on Earth, initially to adapt to the pressures of a still-growing population, shrinking natural resources and climate change. At first, such changes will seem more connected with seemingly prosaic and grounded activities such as producing food, energy security or town planning than with space travel. The future may show such developments to have been useful preparation.

First, humans will need to develop new kinds of agriculture to feed a population that is still growing, to get the maximum amount of nutrition from a minimum of space. To understand this, I shall spend a little time talking about how, in the 1950s and 1960s, *Homo sapiens* used agricultural technology to evade the imminent scythe predicted by Ehrlich in *The Population Bomb*, and how it now needs to step up its game still further. Given that humans ultimately depend on plants for their existence, and currently sequester a large fraction of all the productivity of plants on the planet, I shall dive into the details of photosynthesis, the process whereby plants use sunshine, water and carbon dioxide to create food. It turns out that photosynthesis is a remarkably inefficient process, and even now humans are trying to find ways to improve it – or to develop ways to mimic it artificially. The human future in space will depend on this.

Second, humans must continue to empower women, a development that has occurred only recently in human history, but which must be pursued.

Third, and finally, humans must make concerted efforts to refine and deploy the technologies required to colonize space. After a long hiatus there is currently the germ of a renaissance in crewed space exploration.

This solution will come at a cost, and the currency of

that cost is time. In short, there isn't much of it. *Homo sapiens* lives at a crucial time in its history, that precise inflection point when the ever-rising tide of humanity is about to reach its maximum and then start to recede. If humans do not capitalize on the currently vast reservoir of brains that a large population offers, the colonization of space will falter and sputter out. As I have noted before, it takes a civilization of many hundreds of millions, even billions, to nurture great invention. With some projections suggesting that the population might shrink to less than a billion by 2300 – against a picture of depleted rather than increased available resources – the colonization of space must be well advanced within the next century or two before the human population declines to a degree that it can no longer support the technological innovation and creativity necessary to make the move into space. We must start thinking about it now.

We must first realize that humans will become extinct, eventually. In the long term, all species become extinct. The late palaeontologist David Raup reportedly once quipped that at a first approximation, there is no life on Earth, as 99 per cent of the species that have ever existed are no more.

Species become extinct all the time. There were, however, intervals in Earth's history when the intensity of extinctions far exceeded the usual hum and susurrus of species shuffling silently offstage. Five of these so-called mass-extinction events are known to have occurred over the past 540 million years.[4] The most intense happened at the end of the Permian Period, some 250 million years ago, when a series of super volcanic eruptions released toxic gases into the atmosphere, raising the Earth's mean temperature by several degrees. Some 95 per cent of species in the sea and over 70 per cent of species on land were wiped out over a period of a few hundred thousand years.[5] The next most intense happened sixty-six million years ago, at the end of the Cretaceous Period. This is more famous than the end-Permian event, partly because of its spectacular nature – the wave of extinction was sparked off suddenly when an asteroid hit the Earth – and because the event wiped out the Earth's most celebrated prehistoric residents, the dinosaurs. Today, the human domination of the Earth's ecosystems has raised the possibility that a sixth major mass extinction is currently in progress. The current consensus is that the effects of *Homo sapiens* on the Earth's biological diversity do not match the scale of one of the so-called Big Five – at least, not yet.[6] Humans would have to keep doing what they are doing for another five hundred years for this to register.[7] Yet, as I have documented earlier, the effects of humans on the Earth's biological diversity are already extreme, and I shall discuss this a little more in this chapter.

Now, to imminence. Species of animals and plants tend to have characteristic lifespans. Species of mammal tend to last about a million years, although this varies hugely.[8] *Homo sapiens* is, however, exceptional, in that it is, as far as is known, the only species that has ever evolved on Earth that is conscious of its place in the Universe, and thus able, to an extent, to take charge of its own fate. The Ghost of Christmas Future warns every species of its fate, but only

*Homo sapiens* has the capacity to hear it. But will it listen? To put it another way, can *Homo sapiens* use its own ingenuity to get out of trouble, and beat the odds?

The answer is 'yes' – because it already has, more than once.

For 97 per cent or so of its existence, *Homo sapiens* existed as bands of hunter-gatherers. Hunting and gathering sufficient resources to support people, even to a subsistence level, requires a large amount of territory. Not everywhere is equally suitable – hunter-gatherers tend to live in the places where there are animals and plants to hunt and gather but avoid those areas – often the most biologically rich – where they are likely to fall afoul of diseases.[9] In addition, people could not afford to live too close to one another, because there were not enough resources to support a dense population subsisting off the land. Another way of saying this is that the carrying capacity of the Earth was quite low. One study suggests that the Earth could support ten million humans, if all humans were hunter-gatherers,[10] though estimates vary widely. To a hunter-gatherer, the idea that the Earth would be able to support any more than that would have been ridiculous – and yet the Earth today supports more than eight billion souls.

This transformation came about because of agriculture, a revolutionary lifestyle in which hunters and gatherers stayed at home and cultivated all they needed in a relatively small area. After agriculture was invented, around 10,000 years ago, the human population boomed. Firstly, because agriculture took the ceiling off the earlier restrictive carrying capacity imposed by the hunter-gatherer lifestyle. Second, because, in the process, people domesticated various species of animals and plants to suit their needs – selecting traits such as bigger ears of corn, or sheep with woollier fleece – and in the process creating creatures that produced much higher yields than their wild relatives. Third, because it allowed women to become pregnant more often. In hunter-gatherer societies, where people are always on the move, women space out their pregnancies so that one child is weaned, and mobile, before they have another baby. The relatively sedentary life that went with agriculture lifted that restriction.

And the population boomed, and kept on booming, tracing an ever-upward trajectory until the 1960s, after which the rate of increase started to fall, as we have seen. The Earth held a little more than 3.5 billion people when Paul Ehrlich wrote *The Population Bomb*. Given that most of the Earth's land surface suitable for agriculture was already under cultivation, with the population still growing, Ehrlich and others were worried that the Earth was close to carrying capacity, and without strenuous efforts to reduce population growth, widespread famine would soon follow.[11] And so, like a hunter-gatherer yet ignorant of the promise of agriculture, Ehrlich could do no other than see the human population growing without limit, and to prophesy disaster. Yet amid the gloom there was a glimmer of salvation, even then. 'In my opinion,' he writes, 'the current program with highest potential for reducing the scale of the coming famines involves the development and distribution of new high-yield varieties of food grains.'[12] He goes on to discuss new varieties of rice, corn and wheat that had the capacity to yield far more than the varieties then in use. This was the harbinger of what came to be known as the Green Revolution.[13]

The Green Revolution was arguably the single biggest innovation in agriculture since the invention of agriculture itself. Its effects were immediate. Over the past half century, it has lifted the carrying capacity of the Earth, allowing more than eight billion people to live on our planet's face, and generally – with notable exceptions – in greater affluence than they did in the 1960s.

The raw facts of the Green Revolution are astonishing. For example, it took 10,000 years – since the dawn of agriculture itself – for global production of food grains to reach 1.1 billion tons, a figure reached in 1960. That figure had doubled by 2000, in a space of just forty years, and all because of the Green Revolution. The populations of many low-income countries doubled during that period – but crop yields exceeded even this, reaching an increase of 125 per cent. During this frenetic period, the production of rice soared ninefold (from 220 to 1,984 million tons). Corn and wheat production tripled (from 220 to 660 million tons in each case) – with only a 30 per cent increase in the area of land under cultivation. It has been estimated that without the Green Revolution, food production in developing countries would have been lower by a fifth, while requiring more land; world food prices would have been up to 65 per cent higher; and the average number of raw calories available to a growing population would have declined by as much as 13 per cent.

The Green Revolution began in the 1950s and early 1960s in two places. One was the International Rice Research Institute (IRRI) in the Philippines; the other was the International Corn and Wheat Improvement Center (CIMMYT) in Mexico. The focus on these three species – rice, corn and wheat – was apposite. Together, they account for almost half the calories consumed by *Homo sapiens*.

The techniques used at IRRI and CIMMYT were, in themselves, unremarkable. The scientists at these institutes used conventional plant-breeding techniques – essentially those used by farmers since time immemorial – to cross existing varieties of cereal to produce hybrid strains that would yield more; be more resistant to common pests and diseases; and be suitable for a wider range of more challenging conditions, such as flooding, drought, or acidic or toxic soils previously unsuitable for agriculture. But because they took a scientific, rational approach, the results were achieved far more quickly and with more success than the trial-and-error techniques of traditional farmers unschooled in genetics and formal plant-breeding research.

Before the Green Revolution, rice and wheat plants tended to be tall and spindly with lots of droopy leaves. Applying fertilizer sent them shooting up, so they outgrew their own strength and fell over. The solution was to breed shorter plants with stronger stems. Shorter plants meant more grain per unit of dry weight, so less of the plant's energy was spent on growing long stems that would just end up as waste. The new strains had fewer leaves, but they were darker and erect rather than droopy, maximizing their exposure to nourishing sunlight. The new, improved cereals could be planted closer together, and they grew faster. Rice farmers could now grow two crops a year. Wheat farmers found that the new strains used less water and outgrew weeds, reducing herbicide costs.

For all its undeniable success, the impact of the Green Revolution has been uneven. Whereas the percentage of cropland in Asia planted with the new varieties had reached 82 per cent by 1998, the take-up in sub-Saharan Africa was much lower – just 27 per cent – although this figure has improved since. This was in part because of less interest, at least initially, in crops preferred by African farmers, such as millet, sorghum and cassava. In Asia, the effects of the Green Revolution were felt more in highly irrigated areas than in more marginal areas that depended on rainfall, leading to widening inequity. In general, the benefits of the Green Revolution were felt more by male farmers than by female ones, whose needs were often overlooked. Although fewer people, as a proportion of the world population, went hungry after the Green Revolution, the quality of people's diets did not always improve, especially for the poor. Concentration on growing staples led to a neglect of traditional crops, such as legumes and pulses, that supply essential nutrients and add diversity to the diet. Intensive farming of improved varieties of rice led to a reduction in other beneficial products of rice paddies, such as vegetables and fish.

The Green Revolution has also had environmental consequences. On the plus side, intensive agriculture has reduced the need to bring more land under the plough and has even led to the return of lower-quality farmland to other uses, such as forestry. At the same time, the intensive use of agricultural land has led to soil erosion, pollution and the overuse of increasingly scarce water.

## THE FUTURE IS GREEN AND FEMALE

To be sure, the Green Revolution averted the crises foreseen by Ehrlich and others. Yet in some ways it has been a victim of its own success. The world population continues to increase, and at the same time has become used to higher living standards, putting more pressure on the relatively small number of crops on which humanity depends. Could it be that we have come full circle, with a planet as apparently full to capacity at over eight billion as it seemed to Paul Ehrlich in 1968 when the total was less than half that number?

Estimates of the Earth's current carrying capacity depend rather sensitively on the ground rules you set.[14] In 1679, for example, Dutch pioneer microscopist Antonie van Leeuwenhoek calculated that the Earth could support 13.4 billion people, assuming a global population density equivalent to that of the Netherlands at the time. The number you come up with will depend on a variety of factors, such as quality of life. Clearly, the Earth could support vastly more people in squalor than in affluence. 'The earth's capacity to support people is determined partly by processes that the social and natural sciences have yet to understand, partly by choices that we and our descendants have yet to make,' wrote ecologist Joel E. Cohen in an influential essay.[15]

Another, less direct way to assess carrying capacity is to identify the limits of human activities which, if they are transgressed, will render the Earth much less habitable. This was the approach adopted by Johan Rockström of Stockholm University and colleagues in a 2009 discussion paper.[16] The researchers identified upper limits or 'planetary boundaries' for human interference in nine Earth system processes, including climate change, the rate of species extinctions, disruption to the nitrogen cycle, acidification of the oceans, global use of fresh water and so on. Some of these limits were arbitrary, but the researchers showed that human activities are putting many of these system processes under severe strain and pinpointed three areas in which their limits had been exceeded – climate change (the amount of carbon dioxide in the atmosphere), the rate of species extinction and the amount of nitrogen removed from the atmosphere for human use in chemicals such as fertilisers.[17] As I have shown in earlier chapters, the effects of climate change (to take one example) vary from place to place, and often have disproportionate effects on people without the political power or economic means to mitigate them. A stricter interpretation of planetary boundaries that incorporates an element of what has come to be known as 'climate justice', therefore, shows that almost all planetary boundaries have currently been exceeded.[18] In terms of the disruption that human activity is causing to the normal functioning of the planet, then, it seems that, once again, the Earth has reached carrying capacity.

In some ways we still live like hunter-gatherers. We casually assume that the resources on which we depend are given by the Earth for free, or, at least, at no more cost required than that expended to hunt or gather them. These include access to fresh water and clean air, the formation of soil in which to grow our crops, and the availability of pollinating insects, such as bees, to fertilize them. However,

*Homo sapiens* now dominates the Earth's ecosystems or 'natural capital' to such an extent that many of these so-called ecosystem services have been disrupted by human activity. So much so that it becomes a worthwhile exercise to wonder how much we'd need to pay to replace such services were they to disappear. This was the task of Robert Costanza of the University of Maryland and colleagues, writing in *Nature* in 1997.[19] 'Although it is possible to imagine generating human welfare without natural capital and ecosystem services in artificial "space colonies"', they wrote,

> this possibility is too remote and unlikely to be of much current interest. In fact, one additional way to think about the value of ecosystem services is to determine what it would cost to replicate them in a technologically produced, artificial biosphere. Experience with manned space missions and with Biosphere II in Arizona[20] indicates that this is an exceedingly complex and expensive proposition. Biosphere I (the Earth) is a very efficient, least-cost provider of human life-support services.

Costanza and colleagues calculated that the Earth provides between US$16 trillion and US$54 trillion of ecosystem services per year – an average of US$33 trillion – essentially for free. To put this into perspective, this is around double the global gross national product (as of 1994). Writing about the paper twenty years afterwards,[21] Costanza and colleagues admitted that as a consciousness-raising exercise, their original paper had proven effective and had stimulated debate:

The magnitude of this result [US$33 trillion] shocked some people. Some were surprised because they thought the estimate was too low (an 'underestimate of infinity' by one account), others because they thought it was too high (how could it be larger than the entire world's GDP?), and others because they thought it was a profane and vulgar thing to do in the first place (how can you put a price on nature?). However, most people understood our point in making this admittedly crude estimate: to demonstrate that ecosystem services were much more important to human wellbeing than conventional economic thinking had given them credit for.

The point is this. When the total population of *Homo sapiens* numbered in the low millions, the resources offered by the Earth were effectively inexhaustible and could be taken for granted. Now that people number in the billions – and are straining at or exceeding the Earth's carrying capacity – this is clearly no longer the case. Because the effects of human activity on the Earth's normal functioning are evident, measurable and deleterious, it makes sense to hold these activities to account in a form that policymakers and economists can understand.

It might be no coincidence that the rate of human population growth has started to decline just when the impact of the Green Revolution has reached its limit. It could be that the population has reached a ceiling at which the costs of agricultural improvement now outweigh the benefits, in terms of the erosion of the Earth's natural capital required to sustain humanity. It is possible that new food production technologies[22] will be invented that could raise that ceiling, just as agriculture did 10,000 years ago, and the Green Revolution did from the 1950s and 1960s, but the result might well be a further increase in population until the limits of that technology will be met or exceeded. And let us not forget that the surface of the Earth is finite. No technology will allow the population of *Homo sapiens* to increase without limit.

The coming decline in the human population could be the saving of the Earth, and it is ironic that it was the Green Revolution that made it possible. Because of the Green Revolution, the living standards of many of the world's poorest were raised, and this allowed many of them to do something that they hadn't done before – get an education. There is firm evidence that having an education leads to improved life expectancy, lower family sizes and reduced population growth.[23] The good news is that access to education has risen worldwide over the past few decades.[24] As I mentioned earlier, in 1970, only half of all people in their late twenties had had six years of schooling – that is, they had completed primary education. By 2018, this had risen to four out of five and is projected to reach almost nine in ten by 2030. Access to secondary education has been more mixed. In 1970, around half of all people in their late twenties in the more well-educated parts of the world completed twelve years of schooling, but in many parts of the world the rate was at or below one in ten. The proportion has risen since, but the picture is uneven.

Education has perhaps had a disproportionate impact on women compared with men. After all, education was always more accepting of male than female pupils. That having been said, access to education for girls has improved markedly in recent years. By 2018, almost as many girls as boys had access to education, despite substantial inequalities in Africa and the Middle East. In 1970, men were on average educated to a higher level than women in 142 countries, but the gap has been closing: The figures were twenty-seven countries in 2018, and by 2030 it'll be just four. By then, women are expected to have overtaken men in mean years of schooling in eighteen countries. This represents 'a tremendous reversal of the global landscape that was observed in 1970'.[25]

A consequence of the education of women is to fuel ambition and increase choice. If it is no coincidence that the TFR of most countries is plunging just as the limits of the Earth's capacity have been reached, perhaps, in part, a consequence of the Green Revolution, it might also be no coincidence that we live at the precise moment in history when women are about to overtake men in years of schooling and educational attainment.

It's been a long time coming. For almost the entirety of human history, women had very little agency of any kind. Their function was to produce offspring, which they did as soon as they were able, and continued doing so until they either died in the attempt or reached menopause, whichever came first. The mechanics of birth control were often secretive, a source of shame, illegal, dangerous, or a combination of the above. A turning point was the introduction of the oral contraceptive pill, as recently as the mid-twentieth century.[26] The importance of contraception cannot be overstated. Many projections of world population rely, in some measure, on the provision of contraceptives to women who want or need them.

Against the span of human history, the availability of contraception as well as female education occurred at precisely the same time as female suffrage and power in politics and employment. All may have been benefits, to some degree, of the Green Revolution. Increased food security means that time and resources are freed up for doing things other than growing food, such as going to school. Education and female suffrage will be responsible, to an extent, for the coming downturn in the human population. The downturn will be difficult, and occasionally dangerous, but will be managed in a more civilized and humane way than it might be, thanks to the engagement of women.

## 11

# TURNING OVER A NEW LEAF

*In the revolution of ten centuries, not a single discovery was made to exalt the dignity or promote the happiness of mankind. Not a single idea has been added to the speculative systems of antiquity, and a succession of patient disciples became in their turn the dogmatic teachers of the next servile generation.*

EDWARD GIBBON,
*The Decline and Fall of the Roman Empire*

Thanks in part to the Green Revolution, the burden of the human population is once again straining at the bounds of the Earth's capacity to support it. Although the population will start to decline in the latter part of the twenty-first century, a few more decades of population growth are predicted.

As I discussed earlier, the Green Revolution led to a massive increase in crop yield in the second half of the twentieth century. A lifetime later, however, and with a population more than twice the size it was then, the effects are starting to wear off. For example, between 1987 and 1997 China increased its rice yield from 2.4 to 2.9 tons per acre, but there was no significant increase over the subsequent decade to 2007. Looking at rice yields more broadly, China, India and Indonesia – together the world's largest producers of rice – showed an increase in yield per acre of 36 per cent between 1970 and 1980, but only 7 per cent between 2000 and 2010. In 2008, the world had the lowest stockpiles of wheat for thirty years.

Part of the reason is that conventional plant breeding has gone as far as it can go to improve the yield from the plants we have. It's just not possible to create even bigger, more nutritious grains of wheat or rice, or bigger ears of corn.

Clearly, the Green Revolution has run out of steam. Do we need a new Green Revolution – as some have called it the Green Revolution 2.0 – and, if so, how might this be achieved?

Given that *Homo sapiens* – just one species among millions – appropriates so much of all the products of photosynthesis on the planet;[1] that agriculture drove the massive increase in human population over the past few millennia; and that *Homo sapiens* depends for its continued health and food security on a limited repertoire of crops, it's worth taking some time to explore the inner workings of plants.

In this chapter I take a very close look at photosynthesis, the amazing process whereby green plants turn water, carbon dioxide and sunshine into food, giving off oxygen as a waste product. Plants produce, one way or another, almost all our food, whether we eat them directly, or consume the animals (or their products, such as milk and eggs) that eat the plants. Plants also create the air we breathe. But that's not all. Much of the energy consumed by *Homo sapiens* comes from fossil fuels, which are the products of photosynthesis that happened in the distant past. I do not think it too great a claim that getting to grips with photosynthesis might be vital to the future of *Homo sapiens* – if it has one.

Perhaps the next step will involve digging into the process of photosynthesis itself, to see which parts of it might be improved by direct intervention, producing plants that yield even more than conventional varieties improved by careful breeding.[2] In the process, we might learn so much about photosynthesis that it will be possible to use the energy of sunlight to convert carbon dioxide into foodstuffs entirely artificially, without plants being involved. If this could be achieved, it might be possible to produce food using much less farmland. One solution currently being explored is farming in unusual locations, such as urban rooftops and walls, or vertically stacked hydroponic (that is, soil-free) systems. If food can be created artificially, no farmland need be used at all. Not only would this achieve food security, it would also free up land on Earth for other uses such as forestry, sustainable human habitation or reclaiming wilderness. For thousands of years, agriculture has been tied to land use. Once this link is broken on Earth, sustainable food production in space becomes possible.

Sunshine is the most abundant source of energy immediately accessible to the natural world. Our planet receives about 120,000 terawatts (120,000,000,000,000,000 watts) of energy from the Sun, though it's spread thinly – about 16 watts per square foot per year. For context, humans currently consume about 15 terawatts of energy per year – a minuscule fraction of the total.³ Every year, photosynthesis converts about 220 billion tons of carbon dioxide into food and injects about 154 billion tons of oxygen into the atmosphere.⁴

Although it sounds simple – just mix water and carbon dioxide, add some sunshine, and watch the bounteous food and vital oxygen emerge – photosynthesis is in fact an extraordinarily complex process. This makes it woefully inefficient – only about 1 per cent of the energy harvested by a plant gets converted into food.⁵ The reason for this lies in evolution, which doesn't have to be maximally efficient, only efficient enough to allow any given generation of plants or animals to make it as far as reproducing and producing the next generation. Over the course of evolution, the various deficiencies of photosynthesis have been corrected piecemeal by bodges, kludges and fudges, some of which have introduced their own problems. Evolution is a process of tinkering with what's there, rather than designing something from scratch.

Photosynthesis started early in Earth's history – perhaps as long ago as 3.8 billion years – with bacteria that evolved ways to use the energy from chemical reactions to suck carbon out of the environment, using it to make their own food, in the form of carbohydrates. Later, some bacteria began to evolve pigments that would trap sunlight, using it as an energy source to drive the same process.⁶ The word 'photosynthesis' means 'creation using light'.

Although photosynthesis evolved as many as six different

times in bacteria, the most important, as far as we are concerned, were organisms called cyanobacteria,[7] for two reasons. First, the kind of photosynthesis that evolved in cyanobacteria is directly ancestral to that used in all green plants today, including our major crop plants, and second, because – uniquely – a by-product was molecular oxygen, or $O_2$. All the oxygen we breathe comes from the style of photosynthesis pioneered by cyanobacteria billions of years ago. Once upon a time, cyanobacteria were the dominant creatures on the planet, and the duration of their reign – more than three billion years – makes them arguably the most successful life form that the Earth has ever seen. They are still abundant today. Threads and sheets of cyanobacteria form the blueish-green scum on ponds. Their colour comes from the pigments they use to trap sunlight, driving photosynthesis.

Between 1.5 and 2 billion years ago, more complex life forms emerged from collectives of different bacteria, including cyanobacteria, working together. These collectives became the so-called eukaryotes, made of cells much larger than individual bacteria. All the living things you can see today – all the animals and plants, and you yourself – are eukaryotes, descendants of this original association.[8] Cyanobacteria joined the bacterial collectives and turned, eventually, into chloroplasts. These are the small, bright-green bodies in the cells of plants where photosynthesis happens.

At the heart of photosynthesis in green plants is a chemical reaction in which molecules of water ($H_2O$) are broken into their constituents, oxygen (O) and hydrogen (H). This is a remarkably difficult task. So difficult, in fact, that the trick evolved only once. It is performed by an enzyme called Photosystem II. Enzymes are biological catalysts. That is, they facilitate chemical reactions that would not happen on their own, or if they did, much less often or efficiently. The dismantling of water is one of these.

Strictly speaking, Photosystem II doesn't split water into hydrogen and oxygen. What happens is that four molecules of water are rearranged to make two molecules of molecular oxygen ($O_2$), along with four protons and four electrons, which are sub-atomic particles. Without going into too much detail, a proton is the nucleus of a hydrogen atom when its electron is removed. The electrons and protons released by the reaction drive a whole series of other chemical reactions in a Rube Goldberg–type arrangement of great complexity, the result of which is the accumulation of energy.

Photosystem II is itself a piece of molecular machinery of formidable intricacy that evolved over billions of years into its present form. Even so, it cannot do the job alone. It relies on the participation of another huge molecular machine called Photosystem I. The strange thing is that nearly all photosynthetic bacteria have either Photosystem I or Photosystem II, but not both. The only exceptions are cyanobacteria, and because cyanobacteria are the ancestors of the chloroplasts of green plants, the kind of photosynthesis we see today requires two huge molecular complexes rather than one.

But wait, there's more. The photosystems work closely with further, enormously elaborate arrays of molecular machines called light-harvesting complexes (LHCs). These include pigments such as chlorophyll (responsible for the green colour of plants) that trap the energy from sunlight. Chlorophyll, itself a complex molecule, is not the only photosynthetic pigment. There are many different ones, and it seems that LHCs have evolved many times independently in different kinds of photosynthetic organisms.

This energy accumulated by the photosystems and LHCs fuels the next step: the trapping of carbon dioxide. This is done by yet another big, complicated enzyme catalyst. This one is called ribulose bisphosphate carboxylase, or RuBisCo for short. RuBisCo adds the carbon dioxide to a sugar called ribulose bisphosphate, setting in train another long chain of chemical reactions, the result of which is the production of more sugars, as well as fats and proteins – the stuff from which plants (and we) are made.

So much for the *dramatis personae*: now for the drama.

RuBisCo, like all enzymes, is a kind of protein, and is probably the most abundant protein in nature. The reason for its abundance is that, despite its pivotal importance in nature, any one RuBisCo molecule is spectacularly bad at trapping carbon dioxide, so there must be a lot of them to make up for this deficiency. Up to half of all the protein in a green leaf[9] consists of RuBisCo.

Apart from being rather poor at trapping carbon dioxide, RuBisCo is easily distracted. Instead of picking up carbon dioxide, as it's meant to, it reacts ribulose bisphosphate with any oxygen that's around, a process called 'photorespiration'. Because of this, plants employ a long series of chemical workarounds (some of the bodges, kludges and fudges I mentioned earlier) to minimize photorespiration, by keeping

RuBisCo away from oxygen and bathing it in carbon dioxide instead. These workarounds waste energy that might otherwise be put to useful work. About 30 per cent of the energy plants laboriously hoard by harvesting light is dissipated by photorespiration, and the schemes that plants use to prevent it. All because RuBisCo is so poor at its job.

It gets worse. The mistaken preference of RuBisCo for oxygen rather than carbon dioxide increases when temperatures are high and water is scarce, situations that are likely to increase in frequency as climate change takes hold.

The importance of photorespiration cannot be overstated. To put it into context, just this single process deprives the Corn Belt of the United States of more than 300 trillion calories a year. A reduction in photorespiration of just 5 per cent would translate into more than half a billion dollars extra yield. Is it possible to re-engineer RuBisCo so that it concentrates on carbon dioxide and ignores oxygen?

While we are about it, are there other parts of the machinery of photosynthesis that could be made more efficient by direct intervention?

Half a century on from the Green Revolution, science has not stood still. Back in the 1960s, scientists knew little about how photosynthesis worked, still less how it might be modified. Today, a great deal is known about even the finer details of the process. So much so that it is now possible to create, in a computer, a representation of photosynthesis that allows the easy identification of bottlenecks in the system, parts of the natural photosynthesis that might be tweaked to make it more efficient. That improvement will be thanks to another development, genetic engineering. Introducing genes from different organisms into crop plants – or artificially manipulating the genes already there – might be used to pep up photosynthesis.

One example concerns one of the many enzymes in the photosynthetic machinery, called sedoheptulose-1,7-bisphosphatase, or SBPase. A computer model of photosynthesis predicted that boosting the amount of SBPase in a plant by genetic engineering would, with some other adjustments, boost photosynthetic efficiency by 60 per cent. The prediction bore fruit when applied to experimental crops of tobacco.[10] There are, of course, problems. For example, schemes to reduce or even bypass photorespiration by genetic engineering show promise, but the worry is that there might be unintended consequences – the chemical reactions of photorespiration are linked with other processes in plants that might be vital for their health.

The absorption of light is another issue. Photosynthesis evolved billions of years ago, when the Sun was not as bright as it is today. It also evolved underwater. This means that photosynthetic pigments such as chlorophyll work best in rather low light. In the bright light of a tropical high noon, LHCs receive more light than they can handle and tend to dissipate the excess light as heat, or in driving chemical side reactions that damage the plant. Another factor is that chlorophyll absorbs light only in the visible part of the electromagnetic spectrum. It is insensitive to light in the ultraviolet or infrared range. Even then, it doesn't absorb all the visible light – it doesn't absorb green light. Rather, it reflects it, which is why we see plants as green. There could be scope to engineer new pigments that can respond to brighter light, as well as exploit more of the electromagnetic spectrum than they do currently.

A third problem is the evolutionary accident that plants have two photosystems rather than one. Although both are necessary, they compete with each other as much as they work together, each trying to scavenge the same photons of light from the LHCs. It might be possible to replace Photosystem I with an analogous system from a bacterium. This would come with a different form of chlorophyll that responds to a slightly different range of light wavelengths than the familiar green sort, allowing the plant to use more light, more efficiently.

But the main event is surely the lacklustre performance of RuBisCo, simultaneously the most important and least effective enzyme on the planet. Different plants have slightly different forms of RuBisCo, but it turns out that the ones better at seeking carbon dioxide and shunning oxygen work even more slowly than RuBisCo's customarily near-torpid pace – and that's more lazily than any other enzyme known in the world of nature.

It turns out that modern forms of RuBisCo are optimized for pre-industrial concentrations of atmospheric carbon dioxide. Computer simulations suggest that were it possible to engineer a form of RuBisCo that can work well at the higher concentrations of carbon dioxide found today, the benefit would be 30 per cent more carbon assimilated for the same amount of enzyme. One possibility might then be to grow plants in the flue gases from power plants that burn fossil fuels. These gases typically contain around 10 per cent carbon dioxide, vastly more than there is in the atmosphere (about 0.04 per cent). Such a concentration of carbon dioxide would keep RuBisCo far too busy to be distracted by oxygen, all but eliminating photorespiration. It would also be a clever strategy for recapturing some of the carbon dioxide from power plants before it gets into the atmosphere.

Some plants have evolved a way to get round some of RuBisCo's problems. They do this by absorbing carbon dioxide by other means, and then transporting it to specialized cells in which RuBisCo is confined, releasing highly concentrated carbon dioxide for RuBisCo to work on. These so-called C4 plants include many tropical grasses, and a few crop plants such as corn, sugar cane and sorghum. Indeed, the need to make RuBisCo work better, and to avoid photorespiration at all costs, means that the C4 strategy has evolved independently more than sixty times. It so happens that C4 plants have higher rates of photosynthesis than other, so-called C3 plants. There are efforts under way to convert rice (a C3 plant) to the C4 strategy – but this is not trivial. For one thing, C4 plants have a specialized anatomy as well as many genes that C3 plants do not have.

Whatever else might be done, attempts to re-engineer RuBisCo itself seem to have hit a wall. Perhaps this should have been predicted. If it were possible to create a more efficient version of RuBisCo, plants would have evolved it long ago. If C4 plants got round the problem with elaborate schemes to concentrate carbon dioxide around RuBisCo – removing the temptation of oxygen – and have done so on many different occasions, this should have been a sign that re-engineering RuBisCo itself was never likely to work.

But if photosynthesis in plants is the way it is as a result of almost four billion years of bodges, kludges and fudges, one is entitled to ask whether adding yet more bodges, kludges and fudges by genetic engineering will make plants materially more efficient, or just create more problems. Instead, one could ask if it's possible to turn sunlight into food more simply, getting round the need to (say) engineer new forms of RuBisCo, or convert rice into a C4 plant, or any other of the many schemes proposed to turbocharge photosynthesis – all of which have difficulties, many of them seemingly insuperable. And that is to find a way to convert sunshine into food entirely artificially, with no plants being involved at all.

For a start, the technology that allows the conversion of sunlight into electrons is already well established. This is the basis of the photovoltaic cells in solar panels now familiar from any cityscape. The next step is to harness the power of sunlight to convert carbon dioxide into more complex carbon compounds – the basis of food.

Research into artificial photosynthesis is a busy field, and there are many different approaches to the problem. Some researchers wish to use the power of the Sun, and abundant carbon dioxide, to create industrial chemicals that would otherwise be made from fossil fuels.[11] Other systems combine water-splitting catalysts that produce molecular hydrogen ($H_2$), itself a useful fuel. Some bacteria use hydrogen as an energy source, powering their conversion of carbon dioxide into useful chemicals.[12] Another approach uses industrial catalysts to turn carbon dioxide into simple chemicals, using these as feedstock for a custom-designed array of enzymes from a variety of micro-organisms to turn them into starch – a staple carbohydrate – with more than eight times the efficiency of corn, and entirely in the laboratory, with not a plant in sight.[13]

One of the most exciting developments uses photovoltaic cells to drive a two-step catalytic process[14] that converts carbon dioxide into acetate (acetic acid, the basis of vinegar). This can be used as a feedstock for algae, yeast or other fungi, and even plant tissue, that can be cultivated entirely in the dark, and made into food for human consumption.[15] The process has four times the efficiency of natural photosynthesis, and would be suitable, the researchers say, for growing a lot of biomass in a small space on Earth (thus reducing the pressure on land for agriculture). Crucially for what follows in this book, this kind of approach would be ideal for growing food in the confines of spacecraft, where there is a need to remove carbon dioxide exhaled by astronauts and full-spectrum sunlight is available round the clock.

All the above strategies – engineering plants to make better use of photosynthesis, or creating a kind of photosynthesis that does without plants completely – are likely to take many years of development before they can be applied at such a scale that they have an impact on post–Green Revolution food security, getting *Homo sapiens* over the hump of maximum population and into the next century.

An additional problem is one of regulation. It has been possible to modify plants genetically for forty years. Crops genetically modified to resist herbicides and insect pests – crops such as soybeans, cotton, corn and oilseed rape (canola) – are now routinely grown in many countries.

Development of crops to resist the consequences of climate change, such as flooding, drought and rising carbon dioxide concentrations in the atmosphere, are ongoing. However, some countries still resist the introduction of genetically modified crops, and in any event, bringing such crops to market is liable to get bogged down in regulatory quagmires.

There is one solution, however, that can be applied immediately and requires no sophisticated science at all. That is to eliminate the significant amount of food that is wasted or spoiled before it gets to the consumer, and then, to remove all animal products from the diet.

Across the globe, almost a third of all food is lost between farmyard and dinner plate, by leaks in the supply chain and food spoilage. But these losses are dwarfed by that represented by the consumption of animal products, including dairy and eggs. This is because it requires ten times as much farmland to produce edible animal protein than the same mass of plant protein. As we have seen, photosynthesis is an inefficient process, and this inefficiency is magnified when the products of photosynthesis – crops – are used to feed animals, which are then used to feed people.

Beef is by far the most expensive kind of meat in terms of protein lost. Taking into account losses in the supply chain, the amount of land required to raise 3.5 ounces of plant protein produces just 0.14 ounces of edible beef. A recent study estimated that were all Americans to switch to a plant-based diet nutritionally equivalent to meat, the saving in land would be sufficient to feed an additional 350 million Americans.[16] This figure would be much less for societies that predominantly live on plants, in which meat is eaten infrequently. Fortunately, there is much research ongoing on sources of edible protein that do not start with cattle or sheep grazing in fields. Meat can now be grown from cultured cells in laboratories, and meat-like flavours can be obtained by substituting meat with cultured proteins from fungi, algae, bacteria – even insects. Each strategy has its advantages, but at present it is not clear which can be produced at scale any more cheaply (or less expensively) than meat currently is[17] – though protein derived from fungi is probably the front runner.

But the lesson is clear. A simple route to navigate humanity into a secure post–Green Revolution future and then into space would be to wean *Homo sapiens* off the meat-on-the-hoof carnivory that has otherwise stood it in good stead for almost all its evolutionary history. Simple, though not easy. It takes time and effort to alter consumer choice.[18] However, if food security is threatened by the failure to upgrade food production to Green Revolution 2.0, people may find that the choice has already been made for them.

## 12

# EXPANDING THE HUMAN NICHE

*The Romans made war in all climates, and by their excellent discipline were in a great measure preserved in health and vigour. It may be remarked, that man is the only animal which can live and multiply in every country from the equator to the poles. The hog seems to approach the nearest to our species in that privilege.*

EDWARD GIBBON,
*The Decline and Fall of the Roman Empire*

A network of paths winds through the Bolivian Amazon. Some of the paths are raised above the surrounding landscape, like causeways. A few have gaps. To the untrained eye, the network appears random, a product of natural forces. The truth is surprising. The network of paths and causeways is entirely artificial, created by the dryland fishers of Bolivia. Every now and then, the river floods the landscape. As the floods drain away, the water concentrates in puddles and pools, dammed by the causeways. In the gaps are nets and traps, ensnaring any fish trying to get back to the retreating river.[1] When the flood subsides entirely, the pools form oases, rich in useful plants to supplement the diets of the dryland fisherfolk.

This looks like an unusual kind of farming to a person used to the European or North American landscape of tidy fields, or bare hillsides dotted with sheep. We bring our preconceptions to any new landscape: To a Western farmer, the Amazon is an untamed jungle. In truth, the Amazon was once home to a much greater density of population than it is now, living in extended garden cities rich in urban and ceremonial structures,[2] and tilling a rich earth. The same was true for many other parts of the tropical Americas. When the conquistadores arrived at the end of the fifteenth century, their diseases were heralds of their arrival. By the time Western eyes first looked on the Amazon, the ancient civilizations of the jungle had all but disappeared, reclaimed by forest. So, what we think of as the pristine wilderness of Amazonia is anything but – it is a product of human activities stretching back more than 12,000 years, to when *Homo sapiens* first came to this part of the world.

This realization forces one to look upon one's own environment with fresh eyes. That landscape of tidy fields and bare hillsides dotted with sheep is artificial, too, an entirely human-generated scene, as industrial, in its way, as strip malls and smokestacks. Campaigners who object to new developments such as wind or solar farms on the grounds that they spoil the beauty of the countryside often forget that the countryside we affect to love so much is far from natural. It is entirely human-made and can be maintained as such only by ceaseless vigilance. Left to itself, the trees would reclaim it. Ten thousand years ago, Europe and eastern North America were largely covered in forest.

Today, with agriculture in retreat, these parts of the world are becoming forested again. Early colonial eastern North America was a patchwork of fields and farms, many abandoned to forest as pioneers went west. Today, 7.9 million acres of the United Kingdom – about 13 per cent of the land area – is covered in forest. This is almost three times as much as in 1905.[3] Before one gets a picture of Britons being unusually fond of trees, the picture is the same worldwide. Between 1982 and 2016, global tree cover increased by 865,000 square miles – about nine times the land area of the United Kingdom. Deforestation in the tropics gets all the headlines but is more than matched by reforestation in temperate regions. Most of the change can be put down to a single factor – human activity.[4]

And so, it has always been. Landscape never stands still, and since the spread of *Homo sapiens* around the world, there is probably no longer any landscape that can be regarded as entirely pristine.[5] Human activity even reaches parts of the world rarely, if ever, visited by humans. The strandlines of even the remotest uninhabited islands in the Pacific Ocean are decorated (if that's the word) with human refuse.[6]

Animals and plants modify the environments in which they live, simply because they exist. In ecological terms, each species creates, or constructs, its own niche. These niches can be tiny, or they can be enormous. An example of the small: Fig wasps live out much of their lives inside special structures they coax from fig trees. These are the structures we think of as figs, the fruit of the tree.[7] These insects can be said to have constructed their own ecological niche, something that wasn't there before.

At the opposite end of the scale, much of Eurasia during the last ice age was clothed in an ecosystem called 'mammoth steppe'.[8] This environment supported a rich diversity of plants – mostly grasses and herbs – the like of which is not seen in any modern ecosystem. Grazing the abundant lawn was an almost unimaginable tonnage of large animals, not just the woolly mammoth, as discussed in Chapter 5.[9] Climate change at the end of the ice age withered the mammoth steppe, and, with it, the animals that depended on it (though humans no doubt hastened the process). The animals and plants depended on one another – the plants were nourished from the droppings of the animals; the animals were fed from the rich fodder. Each carved out a niche from the other.

Humans, too, are a part of nature, and have created their own niche, but we have taken it to another level. What is most surprising about human niche construction is that it has been going on for much of our history – perhaps far longer than people who look at the Amazon and see unblemished rainforest realize. Environments have been modified by human action for many thousands, if not millions, of years.[10] Humanity's ancestor *Homo erectus* was the first organism that we know of that could tame and use fire. There are signs, for example, that early humans used fire to clear forest, whether to flush out game in order to hunt it; to entice game to the fresh plant growth in forest clearings; or to promote the growth of plants useful to themselves. The disappearance of the megafauna at the end of the last ice age is mainly attributable to human action. This would have had profound effects on the entire ecosystem, affecting everything from the persistence of fire to the dispersal of fruits and seeds. It would even have affected the climate. For example, the disappearance of large plant-eating animals by overhunting led to the growth of forests, which are better than bare ground at absorbing solar energy. Even in prehistoric times, then, the activities of humans conspired to change the climate.

The invention of agriculture ramped up the effects of the human niche. Deforestation and ploughing release carbon dioxide, and the cultivation of rice led to an increase in the emission of methane to the atmosphere. Methane, even more than carbon dioxide, is a potent greenhouse gas. But what agriculture has done, perhaps more than anything else, is to move animals and plants around the globe, often far from their native ranges. Not just domestic animals, but pests and pathogens, too. We have become accustomed to this migration to such an extent that we assume that animals and plants alien to a landscape must have been there for all time. Rabbits, so common in the English countryside, were introduced – along with many other animals and plants – by the Romans. Once, having breakfast on a terrace at a hotel in Hawaii, I looked around at the gardens and realized that every one of the plants and birds I could see had been introduced: The native flora and fauna of Hawaii are either extinct, or confined to a few relatively remote patches of jungle. Closer to our own times, what would Indian or Chinese cuisine be without the chilli pepper, or the chequered history of Ireland be without the potato? Both are native to the Americas, taken to their new homes within the past 500 years. The net result of the movement of entire suites of animals and plants has been to make landscapes more habitable for humans, thus expanding the human niche.

Islands have been especially strongly affected by what the archaeologist Nicole Boivin and colleagues call 'transported landscapes'.[11] Before humans arrived on the island of Cyprus, for example, there was little to live on. The early farmers who made landfall there brought what they needed to survive, as a complete package. Not just all the crop plants and domestic animals, but wild species too, such as deer, wild boar and foxes. Similar tales can be told for islands across the world that would never have been able to support a human population but for the animals and plants the colonists brought with them.

The effect of humans – and the human introduction of new species to places that they had not previously lived – has been immensely damaging to local biodiversity. Biodiversity loss, like deforestation, grabs all the headlines. It's perhaps surprising to learn, then, that human disturbance might actually have increased biodiversity.

In ecology there is a principle called the 'intermediate disturbance hypothesis'. When ecosystems are left to themselves, developing entirely undisturbed by external forces, what happens is that at most a few species come to dominate, reducing biodiversity. Biodiversity is also reduced at the other extreme, when an ecosystem is disrupted by some vast catastrophe, whether natural (a volcanic eruption, an asteroid impact) or caused by human activity (paving paradise to put up a parking lot).[12] But when the degree of disturbance is somewhere in the middle, between these two extremes, biological diversity blooms. Think of what happens in a forest when an old tree falls over and dies.

The resulting sunny space is soon colonized by a wealth of plants and animals that would not survive beneath unbroken forest cover. The carcass of the old tree decays, providing homes for a multitude of insects, fungi and other organisms. *Homo sapiens*, by shaking up the landscape, has created patchworks of different habitats that would never have formed on their own, and forced different species to come into contact that might not otherwise have done so, the net result being that biodiversity, by some measures, has increased.[13]

Humans have been shaking things up since well before recorded history, creating a niche for themselves that now encompasses the entire planet. As Boivin and colleagues say, '"Pristine" landscapes simply do not exist and, in most cases, have not existed for millennia. Most landscapes are palimpsests shaped by repeated episodes of human activity over multiple millennia.' Human niche construction has been, and continues to be, 'a major evolutionary force on the planet'.

All this human-engendered disruption has taken place against a backdrop of a climate that for the past 6,000 years or so – encompassing the entire span of recorded history – has been relatively stable. *Homo sapiens* has become accustomed to this stability. Perhaps a little *too* comfortable.

During this period, humans seem to have become used to a niche that is much narrower than generally supposed. In general, people tend to live in places with a mean annual temperature of between 51.8 and 59 °F. The exception is the Indian monsoon region, where people are concentrated in places with mean annual temperatures of between 68 and 77 °F. Current projections of climate change show that this temperature niche is likely to shift more in the next fifty years than in the past six millennia. A third of the population will experience mean annual temperatures of greater than 84 °F, a situation currently found on less than 1 per cent of the Earth's surface – mostly in the Sahara Desert.[14] The human response to this change will be to either migrate, or adapt, or both.[15]

It's worth noting, however, that the current interval of stable climate is highly unusual. For almost the entire span of the history of *Homo sapiens*, the climate has alternated, sometimes at breathtaking speed, between long periods of intense cold and shorter episodes of warmth, sometimes tropical warmth, even at high latitudes. Humans have taken these changes in their stride, by both migration and adaptation, and, crucially, by expanding the human niche.

This niche started to expand the moment that one of our australopithecine ancestors first smashed two rocks together to find that the sharp edges of the resulting fragments made useful tools – to dig, to cut and to slice. Even before the advent of cooking, hominins discovered that pounding fibrous plant matter with rocks, or using rocks to smash animal bones to expose the nutritious marrow,[16] expanded their horizons – nutritionally, conceptually and technologically. The discovery of fire was a game changer. Not only could humans begin to cook food – releasing more nutrition and killing parasites – they could use fire to harden stone edges, use the heat to experiment with chemistry (it is possible that Neanderthals used fire to create a useful tar-like adhesive from birch wood)[17] and, most of all, alter the landscape in which they lived. Long before the invention of agriculture,[18] fire was used to burn huge swathes of landscape to flush out game, for example.

Humans started out as opportunist scavengers and predators of the tropical savannah but expanded their niche into colder regions with the help of fire, artificial shelters and clothing. The invention of clothing, incidentally, created an entirely new niche for a parasite – as I described earlier, the human body louse depends entirely on human clothes, and evolved from the head louse at about the time that humans invented *couture*.[19] Humans became adapted to living in all environments, from the seemingly barren Arctic to the complex ecosystems of tropical rainforests, both of which would have been off limits to their savannah ancestors – and all in a geological eyeblink. Although the past six millennia have been relatively stable in terms of climate, humans have used this time to create entirely new and thoroughly artificial habitats. These habitats, which we call 'cities', are, increasingly, places with their own climate, which can to an increasing extent be controlled by their inhabitants. These are the habitats in which most humans now live.[20] But the impact of humans is felt everywhere. From cities to rainforests, from the coasts to high mountains, the human niche now encompasses the whole planet.

So now I come to the point of the past few chapters. Humans have a chance to avoid extinction by expanding their niche still further. They can do this by moving into space. But they will have to do it soon – the launch window is narrow, and it is closing. The sharp fall in the human population in the next century or two will put severe strain on the technological innovation necessary to make this happen – technologies as varied as learning to create genuinely sealed, self-sufficient ecosystems, to the artificial photosynthesis required to make habitats in space liveable, to the mechanics of moving sizeable celestial bodies. The required technologies are in their infancy and will need the application of considerable human ingenuity to raise them to maturity. It's worth repeating the mantra that whereas it takes a village to raise a child, it takes a civilization of billions to create an Einstein. On the other hand, when humans put their minds to it, they can develop technology extraordinarily quickly. Humans progressed from the first powered, controlled, steerable airplanes (1903) to the first Moon landing (1969) in less than a human lifetime – though that development took place against a background of rapidly accelerating population growth. More recently, the responses to the COVID pandemic were innovation in medicine and a greater understanding of vaccination and immunity – in a period of a few months.

In the remainder of this chapter, I shall explore the possible near future of the human niche.

Moviegoers might remember the scene near the beginning of Stanley Kubrick's film *2001: A Space Odyssey* in which the apeman throws the bone into the air – only for its new-found weapon to turn into a spacecraft. The scene is full of meaning about progress and manifest destiny. One interpretation might be that once humans invented technology, the venture into space became all but inevitable.

Here's why.

## EXPANDING THE HUMAN NICHE

A function of technology – some might say the primary function – is to help humans resolve problems posed by their environment with greater elegance, convenience and, more pertinently, economy. An apeman that uses a bone as a club to whack another apeman (as in *2001*) will, because of increased leverage, deliver a far more impactful blow than had it used its bare fists – but for the same amount of energy expended. Clothes, fire and simple shelters made of wood and animal hide not only keep out the cold but reduce the amount of energy a human would need to keep itself warm compared with by, say, shivering. Crucially, such developments allow humans to live in environments where no amount of shivering would suffice to keep them alive.

And that's the key.

Technology allows humans to live in places that might otherwise be closed to them, by shaping the environment to suit their niche. We have seen that humans have changed the Earth to a degree that nowhere on the planet can be regarded as entirely free from human influence. The converse is also true, that is, humans have modified the planet to suit themselves. Although it is true that climate change (itself a side effect of human niche construction) might render parts of the Earth uninhabitable, that presupposes that humans will lack the ingenuity to conquer even these obstacles. In some parts of the world, for example, life in searing heat is made more tolerable by that quintessential twentieth-century invention – air conditioning. It might be said that for those in such parts of the world with sufficient means, life consists of scuttling between one air-conditioned interior and another. The same is true, in reverse, for cities in much colder places, where climate-controlled, snow- and ice-free underground spaces make transport and life in general much easier during the long winter months. By using technology to shut out hostile environments, humans have become expert at living in places far from the tropical savannah in which they evolved. Places such as deserts, the polar regions – and space.

The human journey into space, therefore, does not start with the dramatic roar of powerful engines and the rise, seemingly in slow motion, of a giant space rocket from a launch pad. It begins much closer to home, in the entirely artificial environments in which most humans live out their lives. Most of the animals with which we are familiar[21] – and, for certain, all large mammals – live in the open air, at the mercy of the elements. Except for an unforgivably large and increasing number of humans forced by poverty or refugee status to live in temporary, makeshift shelters at best, humans live indoors, in environments that they themselves create and control.

The lives of wild animals and plants are in many ways determined by the lives and habits of the other species with which they share their environment. The life of a human living in a modern city, in contrast, is almost entirely shaped by the constructions of other humans – from the house or apartment in which they live, to the office or factory in which they work, to the means by which they commute from one place to the other. The same is true of the places in which city dwellers spend their leisure time – from cinemas to gyms, shopping malls to bowling alleys. The artifice is so complete that a single human translocated from its urban environment to the wilderness would find it extremely difficult to create, unaided, any kind of life that is even a fraction as comfortable as the one enjoyed in the city. Such a human would be as helpless as a single ant removed from its nest and the society of its fellows. Exposure to nature or wilderness comes in small, safe doses. And, as I've shown, such wilderness has already been heavily influenced by human action and might itself be artificial. What's more, the homes and workplaces of humans, and the vehicles in which they travel, are increasingly cut off from the outside world. They are climate-controlled; the air is filtered to remove pollen, dust and micro-organisms; even the sunlight through the windows is polarized or filtered to screen out harmful ultraviolet rays. In many ways, humans live their entire lives in the prototypes of spacecraft, all but completely protected from an external environment that could be hostile. If this sounds far-fetched, consider the pilot, crew and passengers of a modern jet aircraft flying at several hundred miles per hour, 30,000 feet above the ground, who survive only by being in a pressurized cabin entirely sealed off from the exterior. More, think of the crew of a submarine travelling at depths in which the unexposed human would not survive for one instant. Spacecraft, then, are extensions of a tendency in niche construction in which humans have contrived to separate themselves from the external environment almost completely.

Not, however, entirely – and that's important.

Journeys in the sealed, pressurized containers that are jet aircraft always come to an end (though for many passengers crammed into economy class, that end cannot come soon enough) and the time comes to disembark. As the tired passengers stream up the jetway, wastes from the lavatories and galleys are removed; the aircraft is cleaned and refuelled; and more freight, passengers, luggage, food and drink are taken aboard for the next journey. The open doors permit the air to be refreshed from outside. In short, airplane flights are finite. Nobody expects the passengers and crew to grow their own food, recycle their own wastes or replenish their own air supplies for a journey of indefinite duration.

The same is true for any journey, including space missions. Each space mission has a set timetable, after which the spacecraft is expected to fail. Even the International Space Station was built in the expectation that it would be decommissioned, someday. To be sure, some uncrewed spacecraft can – and do – remain operational years, perhaps decades after their projected lifetimes. When the spacecraft has human crew, however, the missions are severely constrained. A return to Earth is as much a part of a highly choreographed plan as the initial blast-off into space. As yet there have been no crewed space missions that are either self-sufficient or open-ended.

There have been several attempts to make habitats on Earth for humans that are entirely sealed off from the outside world, with the aim of finding out what problems might be encountered by future colonists of habitats in space. One such is Aquarius, a cylindrical habitat measuring fifty feet long and fifteen feet in diameter (similar in dimensions to the US laboratory module on the International Space Station), anchored 200 feet beneath the sea off Florida. This submerged structure is used to simulate life in space. The 'aquanauts' practise 'spacewalks' in the effectively gravity-free environment of the sea, and rehearse routines and handle technologies that might, one day, be used in space.[22] However, Aquarius receives its air supply from the surface, and missions are, like any space mission, of finite duration.

Perhaps the most famous (or notorious) rehearsal for life in space happened aboard Biosphere 2, a terrarium extending over 3.14 acres (slightly less than the size of two soccer fields) of Arizona desert.[23] Biosphere 2 was conceived in the 1970s by a small group of visionaries. By the time it was ready to receive its first crew, it contained more than 3,000 species (including hummingbirds and bushbabies) spread over a variety of habitats including a tropical rainforest and a miniature ocean, complete with coral reef, as well as areas given over to growing crops. The addition of eight humans – a finishing touch – came on 26 September 1991. The intention was that the humans would be entirely self-sufficient in food, air and water, in principle, indefinitely. Sadly, it was not to be. Bacteria in a soil too rich in organic matter outcompeted the humans for oxygen. Concrete in the walls sucked carbon dioxide out of the air, starving the plant life. The ecosystem was unstable. Bees and hummingbirds, introduced as pollinators, died. Crops failed. The crew, now gasping on a concentration of oxygen normally found at an altitude of 14,000 feet, were only saved by pumping in extra oxygen. They had been receiving other supplies in secret, too – a fact that only became clear much later. Finally, on 4 April 1994, two former crew members, concerned about the welfare of the current crew, broke the seals on the doors and let fresh air in. This experiment in indefinite self-sufficiency met an ignominious end.

Biosphere 2 is now run by the University of Arizona as a test bed for experiments on the effects of climate change on ecosystems. However, the idealistic initial experiment was not quite the failure that it was painted at the time. The founders and crew learned a lot about what it takes to create an entirely sealed, self-sufficient habitat. Mainly, that creating a habitat that remains stable over the long term is extremely challenging – and has not yet been achieved. As for Biosphere 2, the failure was perhaps not that it was too ambitious, but that it was not nearly ambitious enough – perhaps by several orders of magnitude.[24] Just over two acres divided into a range of different ecosystems is probably much too small to be viable, or stable. Before people venture into space as anything more than short-term visitors, there needs to be much more research on the creation of self-sufficient ecosystems that can support human populations – at first a few people, then dozens, then perhaps thousands or even millions.

Humans are most of the way towards discovering such solutions in the increasingly controlled, urban, artificial environments in which they live. It could be that some cities, experiencing the heat, cold or extreme weather events of climate change, and the difficulties of keeping large populations supplied with food and drinking water, will first elect to cover large metropolitan areas in transparent domes, inside which heat and humidity might be moderated and inclement weather kept out. Such cities might then aim for the total internal recycling of water, wastes and air. They might do this with the aid of the same artificial photosynthesis that could be used to generate feedstock for fungi and plants that produce large quantities of useful proteins in relatively small spaces. If so, then the translocation of such cities into space would be the final step. Experiments in the cities in which we live, rather than small habitats such as Aquarius or Biosphere 2, will be key to understanding the factors that will make life sustainable and bearable in space for millions of ordinary people indefinitely, rather than for a few trained astronauts for missions of limited duration. The increasing pace and severity of climate change will force cities to make that choice.

I've already mentioned the International Space Station, a collaborative venture that is home to astronauts for various durations. The first habitats in space were small, fragile space stations in low-Earth orbit, housing only a handful of people at a time. To date there have been about thirty-four different space stations.[25] The first was the Soviet *Salyut 1*, launched as long ago as 1971. *Skylab*, the first space station to be launched by the United States, followed in 1973. The first space station designed for long-term occupancy was the Soviet *Mir*. It took a decade to assemble all its components following the initial launch in 1986. The germ of what became the International Space Station was launched into orbit in 1988. China followed with the *Tiangong* space station. The advantage of space stations is that they are modular and can be put together from many pieces individually carried into orbit. The disadvantage is that blasting off from the Earth's surface – required repeatedly both to build and resupply space stations – is expensive and inefficient, even with partially reusable spacecraft such as the United States' now-defunct Space Shuttle fleet.

Another problem is the health of the residents. Although humans can live in space stations for many months, life in zero gravity leads to muscle wasting and bone loss. And only a small number of people can live in a space station at a time. We are a very long way from sustainable, self-sufficient cities in space.

The next target will be the Moon. Only twelve people have so far set foot on the lunar surface, all astronauts in the Apollo program, all of them white and male, and none since 1972. After a long period of almost supine quiescence, there are at last plans to extend the human presence in space beyond low-Earth orbit. The goal of the Artemis missions, led by the United States but involving a consortium of many different governments and corporations, is,[26] among other things, to set up a permanent crewed base on the Moon and the Lunar Gateway space station in orbit around it. The first (uncrewed) test flight of the Artemis system (Artemis I) took place in 2022, and the system is planned to be fully operational by the end of the decade.

The Lunar Gateway is crucial – it will allow for much more economic and frequent transfers of crew and supplies between space and the lunar surface than having to return to Earth each time, contending with the terrific cost of battling against Earth's gravity. And once the Lunar Gateway and surface base are established, they can be extended. A lunar base might be extended almost indefinitely, by digging tunnels underground to shield human residents from the harsh radiation of space as well as the extremes of temperature found on the lunar surface. The excavated rock can be treated to release useful materials, including – possibly – oxygen and water. Perhaps the astronauts can take with them the things they need to create their own food, such as artificial photosynthesis systems.

Artemis is a US-led enterprise, but it is certain that other countries will be keen to establish independent footholds on the Moon. Colonization of the Moon will then assume a political dimension. Although colonization will not displace any indigenes, there will have to be rules of engagement. Such rules already exist. However, it's worth remembering that the United Nations' extant Outer Space Treaty[27] was drafted before anyone had even set foot on the Moon and could soon start to fray at the edges.[28] At some point in the coming century, there will be polities on the Moon that will seek to become independent of whichever terrestrial nation is sponsoring them. When generations of people start to be born and grow up on the Moon, they will feel little allegiance to nations on a distant planet they see floating in the black, starry sky but have never visited. People growing up on the Moon would never want to visit Earth anyway, with its crushing gravity, over-bright light, the sky that nauseating shade of blue, and horizons so vast as to induce agoraphobia.

After the Moon, the next target is the planet Mars. This is a rather more difficult proposition as Mars is much more remote from Earth, an average of 140 million miles away. Getting there is also extremely dangerous. Thus far, roughly six in ten missions to Mars have ended in failure, and these missions have all been robotic. The technology required to get human beings there safely (and back again) is still under development, though crewed missions to Mars might take place as early as the 2030s.[29] Lessons learned from the Artemis missions – as well as those from nations other than the United

States, and even private corporations – will be vital to make missions to Mars a success. Key to a successful crewed mission to Mars will be the establishment of the Lunar Gateway, or, in fact, any space station around the Moon. This will allow astronauts bound for Mars to pause in lunar orbit and shuttle to Mars much more efficiently than by blasting off from Earth and going there directly.

Once crewed space stations have been placed in orbit around the Moon and perhaps Mars, as well as the Earth, it will occur to some people – initially, the crews required to service the space stations and provide for the well-being of the variously transient population of astronauts – that setting foot on a planetary surface is wasteful, perhaps even unnecessary, particularly when climbing down the gravity well of a planet such as the Earth or Mars (and up again) is so expensive and hazardous. People living for most of their lives on space stations will start to look for accommodation that is similarly freewheeling, if more spacious.

Space-based habitats feature large in science fiction, and there has been no shortage of design concepts for artificial habitats that could support many thousands of people in space indefinitely.[30] Crucial to these designs is some kind of motion to simulate gravity. The wheel-shaped space stations popular in fiction (and, indeed, pictured in *2001: A Space Odyssey*) would be spun at a rate that would give the sensation of weight to residents moving along the inside of the rim. This is the same centripetal force experienced by people on fairground rides or by motorcyclists performing stunts such as the 'wall of death'.

Artificial space habitats that could accommodate large numbers of people indefinitely have several disadvantages. The first is the enormous cost of transporting the large amounts of required material into orbit. The second is providing adequate shielding to protect vulnerable human space colonists from harmful radiation. Both problems might be solved by using ready-made (if not quite ready-to-wear) celestial bodies – the asteroids.

The asteroids are a group of bodies ranging in size from boulders less than 0.6 miles across to small planets. More than 1.3 million have been catalogued.[31] They are mostly found in a wide belt orbiting the Sun between Mars and Jupiter, though a few travel further into space, and the orbits of others take them well into the inner Solar System, where their orbits can cross that of the Earth. Such near-Earth asteroids (NEAs) present a hazard to life on this planet. An asteroid several miles in diameter hit the Earth around sixty-six million years ago, wiping out the dinosaurs and many other living things, and causing ecological disruption that lasted for many tens of thousands of years. For obvious reasons, scientists and policymakers have become quite interested in NEAs,[32] and the trial of technology to nudge potentially threatening asteroids into safer orbits was the object of a recent space mission.[33]

It is a small step – conceptually speaking – from whacking an asteroid with a spacecraft to more refined forms of interaction. It might be possible not just to deflect asteroids into different orbits, but to steer them so they come to rest in some desirable position, such as in orbit around the Earth. However, the technology to move asteroids larger than a few hundred feet in diameter in a usefully short length of time might require the use of technology that does not yet exist, such as rockets powered by nuclear fusion.[34] It is also likely that the residents of Earth would object strenuously were any individual nation, corporation or consortium to start moving asteroids around in near-Earth space, for fear of the kind of calamity that did for the dinosaurs.

Whether they are moved or allowed to remain in their native orbits, asteroids are known to contain a wealth of valuable mineral resources, so the construction of habitats in space might well rely on mining asteroids for raw materials, rather than hauling such resources up from planetary surfaces.[35] And once mined – what then? Rather than discard the hollowed husk of an asteroid, it makes sense to use the cavern so created as accommodation for humans.[36] The inside of an asteroid provides a ready-made habitat, with a rocky surface that would protect humans from radiation from the outside.

How many people could an asteroid habitat accommodate? It rather depends on the size of the asteroid. There are approximately 250 known asteroids that are sixty miles or more in diameter. A back-of-the-envelope calculation shows that a cylindrical cavern, some thirty miles long and six miles wide, inside such an asteroid would yield a surface area on the inside of the cylinder of almost 617 square miles. This is more than twenty-five times the area of Manhattan. At the time of writing, Manhattan has a population density of 71,000 people per square mile,[37] so an asteroid habitat of those dimensions could house more than forty-three million people at such a density – more people than live in the whole of Canada.

Most asteroids are much smaller than this, however. Still, a five-mile-long by one-mile-wide cylindrical cavern in a six-mile-diameter asteroid would have an internal surface area of almost sixteen square miles, a little smaller than Manhattan, and be capable of housing more than a million people at the same population density – the size of a moderate city.

If, in the next century, cities come to be much more enclosed and self-sufficient than they are now, the problem of transplanting the idea of a city from the Earth's surface to a hollowed asteroid should not be insuperable. In fact, people, by then, might have come to think of it as entirely natural.

Hollowed asteroids would have to be rotated along the long axis of the cylindrical cavern to create some kind of artificial gravity, so that people moving around on the inside surface of the cylinder would experience a sensation of weight. The simulated gravity need not be that of Earth. For people used to living in space, the much lesser gravity of Mars or the Moon might do just as well and pose no threat to health. One problem is that many asteroids, having very little gravity themselves, do not have the mechanical strength that would allow them to be rotated at any great rate without them falling to pieces. Even this should pose no obstacle. Once a cavern is created by mining, the inner surface can be sintered or welded to increase strength. And there is even a scheme to deliberately shatter an asteroid, spinning it so fast that it disintegrates. The rubble would be caught by an expanding cylindrical net. The result is a cylinder of compacted asteroid rubble in which a habitat could be constructed.[38]

What might life inside such a habitat be like? The answer is rather like a city, though with a few modifications. Abundant sunlight would be collected by solar panels clothing the asteroid's exterior and piped inside through optical fibres or a system of mirrors, so there would be no shortage of energy. There is no reason why it shouldn't be as bright as day. Oxygen and water would be provided by the same artificial photosynthesis that provides feedstock to the fungi and plants from which food can be grown, fed in turn by carbon dioxide and wastes from humans and animals.

But before we get carried away with thoughts of life in space, it's necessary to list some of the downsides, as Kelly and Zach Weinersmith do in their entertaining and thought-provoking book *A City on Mars*, provocatively subtitled *Can We Settle Space, Should We Settle Space, and Have We Really Thought This Through?*[39] The Weinersmiths argue that for all the technology- and science-fiction-fuelled enthusiasm of would-be space settlers, much more research needs to be done on the biology, medicine, economics and governance of habitats in space before they become a realistic enterprise.

It's notable that only a few humans have ever ventured beyond the Earth's magnetosphere, which protects all life from harmful radiation from space. All were Apollo astronauts, who, as has been noted, were all males in the prime of life, selected carefully for their physical health and technical expertise, and exhaustively trained for the purpose. No woman has ever flown on a Moon mission, irrespective of training, physical fitness or expertise. Neither have any babies, children, senior citizens or people with disabilities or chronic medical conditions, nor, in fact, any regular person on the street. If space habitats are to be a success, they must be able to cope with people in various states of health or illness. The health of the residents will be a top priority. Space medics will have to be prepared for every eventuality – and be able to deal with health problems never before encountered, caused by life in what is, to the human body, an alien environment.

As for reproduction, nobody knows whether it's possible to become pregnant in space, let alone bring a foetus to term, given the harsh radiation environment and variations in gravitational force that space settlers are likely to experience. Nobody knows what abnormalities babies born in space might suffer, whether immediately or in later life. Such potential problems might be countered were reproduction to take place entirely artificially, in a completely controlled environment. At one end of pregnancy, *in vitro* fertilization is now almost routine; at the other, medicine is getting better and better at saving the lives of ever more premature babies. Perhaps, one day, the two techniques will meet in the middle and full exogamy (pregnancy outside the body) will be possible. Leaving aside the various ethical concerns raised by this issue, this prospect seems distant. But settlement in space will be impossible without settlers.

Even if these health and reproductive issues can be solved, space settlements are likely to have small populations, at least to begin with. Earlier in this book I explored the depressing and often tragic consequences of small population size on the viability of populations, and this would apply even more in space than it does on Earth, as the chances of regular infusions of new, fresh genes will be infrequent.

Colonizing space also raises legal and governmental issues that space enthusiasts rarely address, but which the Weinersmiths explore in depth. Earlier I mentioned the United Nations' Outer Space Treaty (OST), which regards outer space as a common good that no polity or individual can claim for itself. It forbids the creation of independent polities in space, even if such bodies could be entirely self-sustaining. At least initially, each space venture would be under the governance of the nation that sponsors or funds it. This applies even if the venture is undertaken by a private corporation or individual. If a tech trillionaire who happens to be a citizen of Ruritania funded and populated an entirely self-sustaining space habitat in orbit around Saturn, the laws and customs of Ruritania and the provisions of the OST would apply. Secession would not be an easy option: partly because the government of Ruritania might object, but also because the secessionists would be fencing off a piece of commons that is meant to be accessible for all humans, not just citizens of Ruritania. Similar legal niceties would also hamper the exploitation for private or national profit of any mineral wealth that might be found in space – in the asteroid belt, for example. There are precedents for this kind of problem, as the Weinersmiths note. The Antarctic Treaty, for example, protects the territorial sanctity of Antarctica. Territorial claims have been made (each one looking like a slice of pizza), some of them overlapping, but these are somewhat dormant given the inhospitability and inaccessibility of that continent. It could be that similar vagueness will apply in space, but as the Weinersmiths note, geopolitical issues in space might end up being fought out on Earth, possibly with nuclear consequences.

Some of us will remember 1982, when Argentina and the United Kingdom went to war over the Falkland Islands (the Malvinas, in Argentina), a small territory in the South Atlantic Ocean, which, although inhabited, has even more remote sub-Antarctic 'dependencies' inhabited only by scientists, and then only occasionally. The Falklands War actually started in South Georgia, a forbidding spot favoured more by penguins than people. Might countries on Earth go to war over disputed claims on the Moon?

Another possible model for a legal framework for use in space is the United Nations Convention on the Law of the Sea (UNCLOS), which does allow for some exploitation in certain circumstances. UNCLOS might be a better model than the Antarctic Treaty for the prosaic reason that Antarctica, for all its harsh environment, exists under a breathable atmosphere, whereas life underwater is similar in many ways to that in outer space. It has been said that *Homo sapiens* knows less about the seabed than about the surface of Mars. Either way, it could be that for space to be colonized safely and with minimal governmental pushback, the governance of Earth would have to be much more united than it is now or is likely to be anytime soon.

All the foregoing raises a big question – *why*?

Why would anyone go to all the trouble and expense of settling in space? Self-determination is likely to be extremely difficult in practice, as well as being a legal minefield. Plundering space for natural resources looks economically inviable in the short term, as mineral wealth, however scarce, is always more profitably extracted on Earth, and will be so for the foreseeable future. An economy based on tourism seems more likely, as well as one based on services that will grow around way stations between the Earth, lunar orbit, and points further out. And despite the success that *Homo sapiens* has had in creating its own niches, space is – one hardly needs reminding – an exceptionally dangerous place.

The reasons given by enthusiasts for why anyone would wish to colonize space are usually much less practical than sentimental. One might colonize space because it is 'awesome', or – as mountaineers explain when asked why they try to scale a high peak, 'because it's there'.[40] Some suggest that the thought of conquering the 'high frontier' is fuelled by some fundamental human urge to explore. Whether this urge is universally human, or in this case informed by American science fiction and myths of the Wild West, is a moot point, though as noted earlier, I remember vividly as a small (British) child reading *You Will Go to the Moon*, with its title loaded, had I but known it, with manifest destiny. And aimed at small children. Whether it is more significant that I remember this particular book so vividly, among many others I must have encountered, or that, this memory notwithstanding, I have not as yet been to the Moon, is for the reader to decide.

There is always the factor of national prestige to consider. This was the major driver of the race between the United States and the (then) Soviet Union to land on the Moon in the 1960s. But that's more about geopolitics on Earth than the conquest of space.

There are, therefore, no rational reasons why anyone

would wish to settle in space. This will, however, not stop people going into space for irrational ones.

Historically, one of the most important reasons for small groups of people settling anywhere far from home is religion. People have moved continents because of religious persecution, or because they want to create an ideal community far from the temptations of humanity in general.

It was religion that prompted many of the early European colonists of North America to cross the wide Atlantic Ocean. They were not driven by the thought of wealth, as were the conquistadores of Central and Southern America, for all that one of the reasons for the English colonization of North America was to interrupt Spanish shipping. Many of the early colonists didn't wait to see if their venture was likely to be practical, still less viable. One thinks of the tragic histories of the early British colonies of Roanoke and Jamestown on the Atlantic coast of what is now the United States. At least as far from home – in terms of time taken to get there – as an astronaut on Mars would be today, such early colonies were repeatedly extinguished by famine and disease, as well as conflict with the indigenous Native Americans. No amount of faith in divine providence could surmount these formidable temporal obstacles. Early settlers in space are highly unlikely to meet hostile indigenes, but the hostility of the environment itself will more than compensate. It took two centuries and a bloody war for England's colonies in North America to become sufficiently well established for them to become a separate polity – the United States of America. The high failure rate of ill-prepared adventurers striking out into the unknown should be a cautionary lesson for would-be space colonists.

It was a mixture of national pride and economic adventurism, rather than religion, that drove the Scottish attempt to colonize part of what is now Panama in the 1690s. The idea was to monopolize trade across the isthmus of Panama, as well as establish a colonial rival to the growing power of Scotland's neighbour, England.[41] The attempt was a total failure, with almost all the colonists dying from disease, famine or conflict with Spain. The venture drove Scotland to the brink of economic ruin. So much so that it had no choice but to subsume itself into the England-dominated United Kingdom, formalized by the Act of Union in 1707. This should be another cautionary tale for would-be space colonists.

I suspect that the human propensity not to listen to advice will be a more pertinent factor in the near-future exploration of space than any supposed lemming-like outward urge that drives *Homo sapiens* as a species. I'd lay reasonable odds that the settlement of space will happen, but it might take several centuries for it to become self-supporting, and there will be untold tragedy and heartache along the way.

The gamble is whether a human future in space can be established in the next two centuries or so, before the population of *Homo sapiens* on Earth becomes too small and sparse to sustain it.

Imagine that it does happen, though – that humans will establish bridgeheads away from the Earth. Whole communities of people – many millions of them, even billions – would grow and develop in asteroid habitats, forming polities as varied as each in religion and custom, and perhaps even evolving new biological characteristics. *Homo sapiens* will branch into a variety of different species, each with its own outlook on the Universe. Some colonies might thrive. Others will surely fail. The culture and politics of some might be anathema to people living in others. In general, though, expansion into space might be the salvation of humanity, ushering in a new age of discovery, exploration and knowledge. Eventually, some habitats will decide to break free from the home Solar System and start to voyage into the wider Galaxy. If so, who knows what the future will have in store?

In his prognostication on the statistics of human extinction, Richard Gott said that a future in space for *Homo sapiens* would be extremely unlikely.[42] But that conclusion did not anticipate where humanity would find itself, poised on a particular cusp. As I hope I have shown in this book, we humans – you, the reader, and I – live at a time that is unique in the entire history of our species, when the population is about to peak and start to decline. It is, therefore, a special time. Things could go two ways. Either *Homo sapiens* will decline and fall to extinction within the next 10,000 years or so. Or it could take space seriously, make a concerted effort to expand into the Universe, and live – potentially – for millions of years.

This is a decision that must be made within the next century, two centuries at most. If *Homo sapiens* is going to take its long-term future seriously, it must start now.

# AFTERWORD

In my book *A (Very) Short History of Life on Earth* I assumed that *Homo sapiens*, like all species, would become extinct – one day. I was, however, rather vague about the timescale, in the end noting that the species would become extinct 'sooner or later'. Extinction is the fate of all species, but *Homo sapiens* is problematic as, *pace* arguments against human exceptionalism of the kind I put forward in an earlier book, *The Accidental Species*, *Homo sapiens* is an exceptional species in many ways, able to modify its circumstances through technology in ways that are unprecedented, as far as we know, in the history of life. With the possible exception of the bacteria that evolved oxygenic photosynthesis more than two billion years ago, releasing the noxious gas $O_2$ into the atmosphere and precipitating the extinction of life that had evolved in its absence – *Homo sapiens* is a uniquely destructive species, both to itself and to the environment in which it lives. This makes predicting human extinction an exercise in uncertainty. Humans are simultaneously amazingly creative and wantonly destructive.

I decided, therefore, to look more closely at the factors that might work against the long-term existence of *Homo sapiens*. The result was an essay in *Scientific American* entitled 'Humans Are Doomed to Go Extinct'.[1] This caused

# AFTERWORD

consternation in many languages. After that, I felt that I should extend my essay into a book, and that's what you see here. I thank Kate Wong for publishing my essay in *Scientific American*; Ehsan Masood and Brian Clegg for taking the time to read drafts of the text; Ravi Mirchandani and Lewis Russell of Picador and George Witte at St. Martin's Press for running with this peculiar project; Mairéad Loftus and her colleagues in the Pan Macmillan Rights department; and my longtime agent Jill Grinberg at Jill Grinberg Literary Management for much encouragement along the way, as well as suggesting that such a depressing topic really ought to have a Hollywood ending. Needless to say, all errors (probably many) and omissions (even more) are my responsibility. As always, I thank Penny and Aviv, and Rachel, who said that if I gave her a special mention in the book she'd make me a cup of tea.

# Notes

### PROLOGUE

1. Fraser, N., and Henderson, D., *Dawn of the Dinosaurs* (Bloomington: Indiana University Press, 2006).
2. And yes, I know, the birds survived, and birds were descended from dinosaurs, but in the mind of the public, the word 'dinosaur' conjures up *T. rex* rather than turkeys.
3. Benton, M., Scientific methodologies in collision: the history of the study of the extinction of the dinosaurs, *Evolutionary Biology* **24**, 371–400, 1990.
4. Zliobaite, I., *et al.*, Reconciling taxon senescence with the Red Queen's hypothesis, *Nature* **552**, 92–5, 2017; Marshall, C. R., A tip of the hat to evolutionary change, *Nature* **552**, 35–7, 2017.
5. The notion of what constitutes a 'human' is notoriously hard to define. In this book I shall use it in some places to refer to a member of our own species, *Homo sapiens*, and in others to any member of the 'hominins', that is, the group of species to which *Homo sapiens* belongs, which is any species more closely related to *Homo sapiens* than to its closest living relative, the chimpanzee. The notions are confusing because *Homo sapiens* is the only hominin still extant. There were many other species (the number varies according to the person doing the counting). In the pages of this book, the meaning of the word 'human' will invariably depend on the context. Like pornography or jazz, the meaning of the word 'human' can be hard to define, though you'll know it when you see it.
6. For robust opposition to this view, please consult my book *The Accidental Species: Misunderstandings of Human Evolution* (Chicago: University of Chicago Press, 2013).

## NOTES

7 With the possible exception of those bacteria which, more than 2.5 billion years ago, evolved a chemical reaction that produced, as a waste product, one of the deadliest gases known, thus setting in train the first and arguably the deadliest wave of extinctions the Earth had seen, or would see. The name of that gas was molecular oxygen.

8 The formal, 'Latin' name for Neanderthals is either *Homo neanderthalensis* or *Homo sapiens neanderthalensis*, depending on who you ask. Throughout this book I shall simply refer to them by the informal 'Neanderthals'.

9 The Denisovans have no 'formal' or 'Latin' name, because they are not known from sufficiently distinctive fossil evidence. If they did, it would probably be something like *Homo altaiensis*. At this point I shall let you in on something I probably shouldn't, but as it's in an endnote that nobody will read, I can tell you. Just you. Let's call it Our Little Secret. As I describe later in this book, the Denisovans are mostly known from the complete DNA sequence extracted from a single nondescript finger bone by Svante Pääbo – pioneer of ancient DNA research – and his colleagues. No, that's not the secret. The secret part is that he and I were discussing the name *Homo altaiensis* over lunch at a Thai restaurant in South Kensington, London, opposite the Natural History Museum, where he had given a lecture. Don't look for it; it's not there anymore. The restaurant, that is. At the time of writing, the Natural History Museum still stands. But I digress. We decided that the Denisovan remains couldn't really have a 'formal' or 'Latin' name as, given that so few actual bones had been found, it would be hard to know one if you found one unless you had access to a sophisticated laboratory. That situation might change once more Denisovan bones are found. Svante's own account in his excellent book *Neanderthal Man: In Search of Lost Genomes*, in which I play a walk-on part, might differ in some respects from my own recollection, although both of us are right. Svante went on to win a Nobel Prize for his work on extracting ancient DNA. I once won a cake in a raffle. Make of that what you will.

10 I apologise to any reader who identifies as non-human.

11 Gibbon published his work in six volumes between 1776 and

1788. My edition is the splendid eight-volume set produced by the Folio Society between 1983 and 1990 and edited by Betty Radice and Felipe Fernández-Armesto.

12 The Roman Empire reached its greatest extent in 117 CE, when Trajan conquered Mesopotamia, which was lost soon after. After that it was downhill all the way. In her book *SPQR: A History of Ancient Rome*, historian Mary Beard places the start of the decline of Rome in the 212 CE edict of the Emperor Caracalla (reigned 198–217 CE) that granted the hitherto prized status of Roman citizenship to (almost) every freeborn man living in the Empire. If everybody's somebody, then no one's anybody.

13 A reminder from before: A hominin is any species more closely related to modern humans than to chimpanzees. All hominins aside from *Homo sapiens* have become extinct.

14 For a full treatment of bipedality, see *First Steps: How Upright Walking Made Us Human* by Jeremy DeSilva (London: HarperCollins, 2021).

15 Hu, W., *et al.*, Genomic inference of a severe human bottleneck during the Early to Middle Pleistocene transition, *Science* **381**, 979–84, 2023.

16 Kaessmann, H., *et al.*, Great ape DNA sequences reveal a reduced diversity and an expansion in humans, *Nature Genetics* **27**, 155–6, 2001.

17 Ragsdale, A. P., *et al.*, A weakly structured stem for human origins in Africa, *Nature* **617**, 755–63, 2023.

18 For a comprehensive treatment of the extinctions at the end of the last Ice Age – many of which can be blamed on *Homo sapiens* – see Anthony J. Stuart's book *Vanished Giants* (Chicago: University of Chicago Press, 2021).

19 Le Page, M., Bananas threatened by devastating fungus given temporary resistance, *New Scientist*, 21 September 2022.

20 Harper, K., *Plagues Upon the Earth* (Princeton: Princeton University Press, 2021).

21 http://www.porphyria-professionals.uct.ac.za/ppb/porphyrias/vp

22 https://psychnews.psychiatryonline.org/doi/full/10.1176/pn.4 0.24.0021.

23 https://www.healthline.com/health/crohns-disease/jewish-ancestry.

## NOTES

24  Lynch, M., Mutation and human exceptionalism: our future genetic load, *Genetics* **202**, 869–75, 2016.
25  In 2022, 9.56 million people were born in China, against 10.41 million deaths. Stevenson, A., and Wang, Z., 'China population falls, heralding a demographic crisis', *New York Times*, 16 January 2023, https://www.nytimes.com/2023/01/16/business/china-birth-rate.html.
26  Ehrlich, P., *The Population Bomb* (New York: Ballantine Books, 1968).
27  Vollset, S. E., *et al.*, Fertility, mortality, migration, and population scenarios for 195 countries and territories from 2017 to 2100: a forecasting analysis for the Global Burden of Disease Study, *The Lancet* **396**, 1285–1306, 2020.
28  Friedman, J., *et al.*, Measuring and forecasting progress towards the education-related SDG targets, *Nature* **580**, 636–9, 2020.
29  Masood, E., *GDP: The World's Most Powerful Formula and Why It Must Now Change* (London: Icon Books, 2021).
30  Vaesen, K., *et al.*, Inbreeding, Allee effects and stochasticity might be sufficient to account for Neanderthal extinction, *PLoS ONE* **14**, e0225117, 2019.
31  Diamond, J., The last people alive, *Nature* **370**, 331–2, 1994.
32  Looked at more broadly, Ehrlich was echoing the doom-mongering of Thomas Malthus who, in his *Essay on the Principle of Population* (first published in 1798), predicted that human populations would naturally outgrow their means of supply. Malthus published six editions of his work. The final one, in 1826, influenced Charles Darwin and offered one of the mechanisms for what came to be known as natural selection. That is, organisms will always produce many more offspring than can possibly survive.
33  Rees, W. E., The human ecology of overshoot: why a major 'population correction' is inevitable, *World* **4**, 509–27, 2023.
34  Krausmann, F., *et al.*, Global human appropriation of net primary production doubled in the 20$^{th}$ century, *Proceedings of the National Academy of Sciences of the United States of America* **110**, 10324–9, 2013.
35  Kolbert, E., Creating a better leaf, *The New Yorker*, 13 December 2021.
36  Parfitt, S., Ashton, N., Lewis, S., *et al.*, Early Pleistocene

NOTES

human occupation at the edge of the boreal zone in northwest Europe, *Nature* **466**, 229–33, 2010; Roberts, A., Grün, R., Early human northerners, *Nature* **466**, 189–90, 2010.
37 See Marshall, T., *The Future of Geography* (London: Elliott and Thompson, 2023).

## 1. THE HUMAN FAMILY

1 Biewener, A., Wilson, A., R. McNeill Alexander *(1934–2016)*, *Nature* **532**, 442, 2016, https://doi.org/10.1038/532442a; Alexander, G., Robert McNeill Alexander, 7 July 1934 – 21 March 2016, *Biographical Memoirs of the Royal Society*, 22 December 2021, https://royalsocietypublishing.org/doi/10.1098/rsbm.2021.0030.
2 Alexander, R., McN., Estimates of speeds of dinosaurs, *Nature* **261**, 129–30, 1976.
3 Leakey, M., and Hay, R., Pliocene footprints in the Laetolil Beds at Laetoli, northern Tanzania, *Nature* **278**, 317–23, 1979; Leakey, M., Hay, R., Curtis, G., *et al.*, Fossil hominids from the Laetolil Beds, *Nature* **262**, 460–6, 1976.
4 I refer to her as 'she' after the most famous specimen of *Australopithecus afarensis*, a partial skeleton of a female discovered at Hadar in Ethiopia and known as 'Lucy'.
5 McNutt, E. J., Hatala, K. G., Miller, C., *et al.*, Footprint evidence of early hominin locomotor diversity at Laetoli, Tanzania, *Nature* **600**, 468–71, 2021.
6 Gierlinski, G. D., *et al.*, Possible hominin footprints from the late Miocene (c. 5.7 Ma) of Crete?, *Proceedings of the Geologists' Association* **128**, 697–710, 2017, http://dx.doi.org/10.1016/j.pgeola.2017.07.006; Kirscher, U., El Atfy, H., Gärtner, A., *et al.*, Age constraints for the Trachilos footprints from Crete, *Scientific Reports* **11**, 19427, 2021. It should be said that not everyone agrees that these footprints were made by a hominin, and some think that they aren't footprints at all.
7 Böhme, M., Spassov, N., Fuss, J., *et al.*, A new Miocene ape and locomotion in the ancestor of great apes and humans, *Nature* **575**, 489–93, 2019.
8 McBrearty, S., and Jablonski, N. G., First fossil chimpanzee, *Nature* **437**, 105–8, 2005.

9 Suwa, G., Kono, R., Katoh, S., *et al.*, A new species of great ape from the late Miocene epoch in Ethiopia, *Nature* **448**, 921–4, 2007.
10 Chaimanee, Y., Suteethorn, V., Jintasakul, P., *et al.*, A new orang-utan relative from the Late Miocene of Thailand, *Nature* **427**, 439–41, 2004.
11 Another orangutan relative, *Gigantopithecus*, from southern China, grew to twice the size of an adult male gorilla, and was the largest primate that ever existed – though its size can only be estimated from its jaws and teeth, for no other parts of its skeleton have been found. The last *Gigantopithecus* died out relatively recently – around 300,000 years ago or so.
12 Brunet, M., Guy, F., Pilbeam, D., *et al.*, A new hominid from the Upper Miocene of Chad, Central Africa, *Nature* **418**, 145–51, 2002.
13 Daver, G., Guy, F., Mackaye, H. T., *et al.*, Postcranial evidence of Late Miocene hominin bipedalism in Chad, *Nature* **609**, 94–100, 2022.
14 Senut, B., *et al.*, First hominid from the Miocene (Lukeino Formation, Kenya), *Comptes Rendus de l'Académie des Sciences Series IIA – Earth and Planetary Science* **332**, 137–44, 2001.
15 Haile-Selassie, Y., Late Miocene hominids from the Middle Awash, Ethiopia, *Nature* **412**, 178–81, 2001.
16 White, T., Suwa, G., and Asfaw, B., *Australopithecus ramidus*, a new species of early hominid from Aramis, Ethiopia, *Nature* **371**, 306–12, 1994.
17 Haile-Selassie, Y., Saylor, B., Deino, A., *et al.*, A new hominin foot from Ethiopia shows multiple Pliocene bipedal adaptations, *Nature* **483**, 565–9, 2012. It's possible that the maker of the tracks at Laetoli that had once been attributed to a bear might have had a foot like this.
18 Kappelman, J., Ketcham, R., Pearce, S., *et al.*, Perimortem fractures in Lucy suggest mortality from fall out of tall tree, *Nature* **537**, 503–7, 2016.
19 With the exception, of course, of kangaroos, but they do it differently: They hop, rather than walk, and use their long muscular tails as a counterbalance.
20 One of the claims of the so-called 'aquatic ape' hypothesis, perhaps the most elaborate scheme to explain human

adaptations after the fact. Other adaptations supposedly connected with aquatic life include a high fat content, the shape of human sinuses, the capacity for babies to swim and relative hairlessness. See Morgan, E., *The Aquatic Ape Hypothesis* (London: Souvenir Press, 1997). Without wishing to denigrate the several enthusiastic adherents of this view, who are of course entitled to their opinions, my copy of this book is filed in my library under 'science fiction and fantasy'.

21 Haslam, M., *et al.*, Primate archaeology evolves, *Nature Ecology & Evolution* **1**, 1431–7, 2017.
22 My mentor R. McNeill Alexander was a pioneer in understanding the mechanism whereby tendons store and release elastic strain energy during locomotion.
23 Jurmain, R., Degenerative joint disease in African great apes: an evolutionary perspective, *Journal of Human Evolution* **39**, 185–203, 2000.
24 Whitcome, K. K., *et al.*, Fetal load and the evolution of lumbar lordosis in humans, *Nature* **450**, 1075–8, 2007.
25 Bluestone, C., and Swarts, J. D., Human evolutionary history: consequences for the pathogenesis of otitis media, *Otolaryngology–Head and Neck Surgery* **143**, 739–44, 2010; Bluestone, C. D., Humans are born too soon: impact on pediatric otolaryngology, *International Journal of Pediatric Otorhinolaryngology* **69**, 1–8, 2005.
26 Esler, M., *et al.*, Consequences of the evolutionary cardiovascular challenge of human bipedalism: orthostatic intolerance syndromes, orthostatic hypertension, *Journal of Hypertension* **27**, 2333–40, 2019.
27 Fay, J. C., Disease consequences of human adaptation, *Applied and Translational Genomics* **2**, 42–7, 2013.
28 Rube Goldberg, to readers in the United States.

## 2. THE GENUS *HOMO*

1 For a comprehensive review on the genus *Australopithecus*, see Alemseged, Z., Reappraising the palaeobiology of *Australopithecus*, *Nature* **617**, 45–54, 2023.
2 Harmand, S., Lewis, J., Feibel, C., *et al.*, 3.3-million-year-old

stone tools from Lomekwi 3, West Turkana, Kenya, *Nature* **521**, 310–15, 2015, https://doi.org/10.1038/nature14464.

3   Zink, K., Lieberman, D., Impact of meat and Lower Palaeolithic food processing techniques on chewing in humans, *Nature* **531**, 500–3, 2016, https://doi.org/10.1038/nature16990.

4   McPherron, S., Alemseged, Z., Marean, C., *et al.*, Evidence for stone-tool-assisted consumption of animal tissues before 3.39 million years ago at Dikika, Ethiopia, *Nature* **466**, 857–60, 2010, https://doi.org/10.1038/nature09248.

5   Villmoare, B., *et al.*, Early *Homo* at 2.8 Ma from Ledi-Geraru, Afar, Ethiopia, *Science* **347**, 1352–5, 2015, https://doi.org/10.1126/science.aaa1343.

6   Spoor, F., Wood, B., and Zonneveld, F., Implications of early hominid labyrinthine morphology for evolution of human bipedal locomotion, *Nature* **369**, 645–8, 1994, https://doi.org/10.1038/369645a0.

7   Brown, F., Harris, J., Leakey, R., *et al.*, Early *Homo erectus* skeleton from west Lake Turkana, Kenya, *Nature* **316**, 788–92, 1985, https://doi.org/10.1038/316788a0.

8   Bramble, D., and Lieberman, D., Endurance running and the evolution of *Homo*, *Nature* **432**, 345–52, 2004, https://doi.org/10.1038/nature03052.

9   Joordens, J. C. A., *et al.*, *Homo erectus* at Trinil on Java used shells for tool production and engraving, *Nature* **518**, 228–31, 2015.

10  Zhu, Z., Hominin occupation of the Chinese Loess Plateau since about 2.1 million years ago, *Nature* **559**, 608–12, 2018.

11  Gabunia, L., and Vekua, A., A Plio-Pleistocene from Dmanisi, East Georgia, Caucasus, *Nature* **373**, 509–12, 1995.

12  Conan Doyle may be long gone, but the Hill House Inn is still there.

13  Parfitt, S., Ashton, N., Lewis, S., *et al.*, Early Pleistocene human occupation at the edge of the boreal zone in northwest Europe, *Nature* **466**, 229–33, 2010; Ashton, N., Lewis, S. G., De Groote, I., Duffy, S. M., Bates, M., Bates, R., *et al.*, Hominin footprints from early Pleistocene deposits at Happisburgh, UK, *PLoS ONE* **9**(2), e88329, 2014, https://doi.org/10.1371/journal.pone.0088329.

14  Carbonell, E., Bermúdez de Castro, J., Parés, J., *et al.*, The

## NOTES

first hominin of Europe, *Nature* **452**, 465–9, 2008; Bermúdez de Castro, J. M., *et al.*, A hominid from the Lower Pleistocene of Atapuerca, Spain: possible ancestor to Neandertals and modern humans, *Science* **276**, 1392–5, 1997, http://doi.org/10.1126/science.276.5317.1392.

15  Arsuaga, J.-L., *et al.*, Three new human skulls from the Sima de los Huesos Middle Pleistocene site in Sierra de Atapuerca, Spain, *Nature* **362**, 534–7, 1993.

16  Jaubert, J., Verheyden, S., Genty, D., *et al.*, Early Neanderthal constructions deep in Bruniquel Cave in southwestern France, *Nature* **534**, 111–14, 2016, https://doi.org/10.1038/nature18291.

17  Skov, L., Peyrégne, S., Popli, D., *et al.*, Genetic insights into the social organization of Neanderthals, *Nature* **610**, 519–25, 2022.

18  Ni, X., *et al.*, Massive cranium from Harbin in northeastern China establishes a new Middle Pleistocene human lineage, *Innovation* **2**, 100130, 2021.

19  Détroit, F., *et al.*, A new species of *Homo* from the Late Pleistocene of the Philippines, *Nature* **568**, 181–6, 2019.

20  Brown, P., *et al.*, A new small-bodied hominin from the Late Pleistocene of Flores, Indonesia, *Nature* **431**, 1055–61, 2004.

21  Grün, R., Pike, A., McDermott, F., *et al.*, Dating the skull from Broken Hill, Zambia, and its position in human evolution, *Nature* **580**, 372–5, 2020, https://doi.org/10.1038/s41586-020-2165-4.

22  Dirks, P. H. G. M., *et al.*, The age of *Homo naledi* and associated sediments in the Rising Star Cave, South Africa, *eLife* **6**, e24231, 2017.

23  Bergström, A., Stringer, C., Hajdinjak, M., *et al.*, Origins of modern human ancestry, *Nature* **590**, 229–37, 2021, https://doi.org/10.1038/s41586-021-03244-5.

24  Reich, D., *et al.*, Genetic history of an archaic hominin group from Denisova Cave in Siberia, *Nature* **468**, 1053–60, 2010.

25  Reich, D., *et al.*, Denisova admixture and the first modern human dispersals into southeast Asia and Oceania, *American Journal of Human Genetics* **89**, 516–28, 2011.

26  Huerta-Sánchez, E., *et al.*, Altitude adaptation in Tibetans caused by introgression of Denisovan-like DNA, *Nature* **512**, 194–7, 2014.

## NOTES

27 Ceballos, F. C., and Álvarez, G., Royal dynasties as human inbreeding laboratories: the Habsburgs, *Heredity* **111**, 114–21, 2013; Álvarez, G., *et al.*, The role of inbreeding in the extinction of a European royal dynasty, *PLoS ONE* **4**, e5174, 2009.
28 Hashmi, M. A., Frequency of consanguinity and its effect on congenital malformation – a hospital based study, *Journal of the Pakistan Medical Association* **47**, 75–8, 1997.
29 Berra, T. M., *et al.*, Was the Darwin/Wedgwood dynasty adversely affected by consanguinity?, *BioScience* **60**, 376–83, 2010.
30 Copeland, S., Sponheimer, M., de Ruiter, D., *et al.*, Strontium isotope evidence for landscape use by early hominins, *Nature* **474**, 76–8, 2011, https://doi.org/10.1038/nature10149.
31 Skov, L., Peyrégne, S., Popli, D., *et al.*, Genetic insights into the social organization of Neanderthals, *Nature* **610**, 519–25, 2022, https://doi.org/10.1038/s41586-022-05283-y.
32 Hublin, J.-J., Ben-Ncer, A., Bailey, S., *et al.*, New fossils from Jebel Irhoud, Morocco and the pan-African origin of *Homo sapiens*, *Nature* **546**, 289–92, 2017, https://doi.org/10.1038/nature22336; Richter, D., Grün, R., Joannes-Boyau, R., *et al.*, The age of the hominin fossils from Jebel Irhoud, Morocco, and the origins of the Middle Stone Age, *Nature* **546**, 293–6, 2017, https://doi.org/10.1038/nature22335.

### 3. LAST AMONG EQUALS

1 For a comprehensive review of the latest thinking on the genetics of the origin of *Homo sapiens*, see Bergström, A., *et al.*, Origins of modern human ancestry, *Nature* **590**, 220–37, 2021. For a more fossil-oriented perspective, see Stringer, C., The origin and evolution of *Homo sapiens*, *Philosophical Transactions of the Royal Society B* **371**, 20150237, 2016.
2 Cann, R. L., Stoneking, M., and Wilson, A. C., Mitochondrial DNA and human evolution, *Nature* **325**, 31–6, 1987.
3 Except for red blood cells – at least in mammals. If your red blood cells contain nuclei, then you are probably a bird.
4 Before I go any further, I should add a very big health warning. Genes aren't really 'sentences', and the sum total of all the genes – the genome – isn't really a 'book', even a book of 'recipes'. Such terms are more or less crude metaphors we

use to describe the interactions of a complex arrangement of molecules performing activities at very small scales, and in physical and chemical conditions that we humans find very hard to appreciate. What I present here, for the sake of time and clarity, is a simplification that verges on the fictional. Genes, made of DNA, do contain instructions for the assembly of molecules of the chemical relative RNA (ribonucleic acid), some of which, in turn, contain instructions for creating proteins. Many other molecules of RNA participate in processes important to the welfare of the cells in which they reside, many of which are, as yet, unclear. Taken together, the activities in and around genes act as starting points for processes which, eventually, result in the development of a human being from a single cell, the merger of a sperm and an egg. Some of these processes are physical, chemical and mechanical, and are not encoded in genes. The wiring of the human brain, for example, is one of these. It is an 'emergent property' of interactions on the scale of molecules but cannot be predicted from them. The development of a human body from a sperm and an egg is similarly emergent and relies on higher-level interactions such as the splitting and merger of various tissues and structures as the embryo develops, none of which are represented as genes. For a new view of the relationship between molecular biology and the processes of life, I encourage every reader to consult *How Life Works* by Philip Ball (Chicago: The University of Chicago Press, 2023).

5   This could very well be a trivial argument. Because the human population has expanded from a very small pool of ancestors, you don't have to be the unfortunate Charles II of Spain to know that the further back you search, the more you'll see the same names cropping up. In *A Brief History of Everyone Who Has Ever Lived*, Adam Rutherford notes that it is almost certain that any person with predominantly British ancestry born in the 1970s will be able to claim descent from Edward III, a king of England in the fourteenth century. All that is lacking is the documentation. Edward III is known to have had many children, many of whom survived to produce many more. (There are Habsburgs, too, alive today, but no person now living can owe any of their DNA to *El Hechizado*.) See

## NOTES

https://www.waterstones.com/blog/family-fortunes-adam-rutherford-on-how-were-all-related-to-royalty.

6 Hublin, J.-J., *et al.*, New fossils from Jebel Irhoud, Morocco and the pan-African origin of *Homo sapiens*, *Nature* **546**, 289–92, 2017.

7 Richter, D., *et al.*, The age of the hominin fossils from Jebel Irhoud, Morocco, and the origins of the Middle Stone Age, *Nature* **546**, 293–6, 2017.

8 Bermúdez de Castro, J. M., *et al.*, A hominid from the Lower Pleistocene of Atapuerca, Spain: possible ancestor to Neandertals and modern humans, *Science* **276**, 1392–5, 1997.

9 Ashton, N., *et al.*, Hominin footprints from Early Pleistocene deposits at Happisburgh, UK, *PLoS ONE*, 2014, https://doi.org/10.1371/journal.pone.0088329.

10 Parfitt, S. A., *et al.*, Early Pleistocene human occupation at the edge of the boreal zone in northwest Europe, *Nature* **466**, 229–33, 2010.

11 McDougall, I., *et al.*, Stratigraphic placement and age of modern humans from Kibish, Ethiopia, *Nature* **433**, 733–6, 2005.

12 Clark, J. D., *et al.*, Stratigraphic, chronological and behavioural contexts of Pleistocene *Homo sapiens* from Middle Awash, Ethiopia, *Nature* **423**, 747–52, 2003.

13 I describe my adventures in my book *In Search of Deep Time* (New York: Free Press, 1999), published in the UK as *Deep Time* (London: Fourth Estate, 2000).

14 Harvati, K., *et al.*, Apidima Cave fossils provide earliest evidence of *Homo sapiens* in Eurasia, *Nature* **571**, 500–4, 2019.

15 McDermott, F., *et al.*, Mass-spectrometric U-series dates for Israeli Neanderthal/early modern hominid sites, *Nature* **363**, 252–5, 1993.

16 Clarkson, C., *et al.*, Human occupation of northern Australia by 65,000 years ago, *Nature* **547**, 306–10, 2017.

17 Westaway, K. E., *et al.*, An early modern human presence in Sumatra 73,000–63,000 years ago, *Nature* **548**, 322–5, 2017.

18 Liu, W., *et al.*, The earliest unequivocally modern humans in southern China, *Nature* **526**, 696–9, 2015.

19 I discuss the issues around ancestry and descent in my book *In Search of Deep Time* (New York: Free Press, 1999), published in the UK as *Deep Time* (London: Fourth Estate, 2000).

## NOTES

20 Krings, M., et al., Neandertal DNA sequences and the origin of modern humans, *Cell* **90**, 19–30, 1997.
21 Reich, D., et al., Genetic history of an archaic hominin group from Denisova Cave in Siberia, *Nature* **468**, 1053–60, 2010.
22 Hublin, J.-J., et al., Initial Upper Palaeolithic *Homo sapiens* from Bacho Kiro Cave, Bulgaria, *Nature* **581**, 299–302, 2020.
23 Fu, Q., Hajdinjak, M., Moldovan, O., et al., An early modern human from Romania with a recent Neanderthal ancestor, *Nature* **524**, 216–19, 2015.
24 Hajdinjak, M., Mafessoni, F., Skov, L., et al., Initial Upper Palaeolithic humans in Europe had recent Neanderthal ancestry, *Nature* **592**, 253–7, 2021, https://doi.org/10.1038/s41586-021-03335-3.
25 Huerta-Sánchez, E., et al., Altitude adaptation in Tibetans caused by introgression of Denisovan-like DNA, *Nature* **512**, 194–7, 2014.
26 Mondal, M., et al., Approximate Bayesian computation with deep learning supports a third archaic introgression into Asia and Oceania, *Nature Communications* **10**, 246, 2019; Mondal, M., et al., Genomic analysis of Andamanese provides insights into ancient human migration into Asia and adaptation, *Nature Genetics* **48**, 1066–72, 2016.
27 Ragsdale, A. P., et al., A weakly structured stem for human origins in Africa, *Nature* **617**, 755–63, 2023.
28 Grün, R., et al., Dating the skull from Broken Hill, Zambia, and its position in human evolution, *Nature* **580**, 372–5, 2020.
29 Crevecoeur, I., et al., Late Stone Age human remains from Ishango (Democratic Republic of Congo): new insights on Late Pleistocene modern human diversity in Africa, *Journal of Human Evolution* **75**, 80–9, 2014.
30 Tryon, C. A., et al., Late Pleistocene age and archaeological context for the hominin calvaria from GvJm-22 (Lukenya Hill, Kenya), *Proceedings of the National Academy of Sciences of the United States of America* **112**, 2682–7, 2015.
31 Crevecoeur, I., The Upper Palaeolithic human remains of Nazlet Khater 2 (Egypt) and past modern human diversity. In Hublin, J.-J., and McPherron, S. (eds.), *Modern Origins: A North African Perspective* (Dordrecht: Springer, 2012).
32 Harvati, K., et al., The Late Stone Age calvaria from Iwo

Eleru, Nigeria: morphology and chronology, *PLoS ONE* **6**, e24024, 2011.
33 Hammer, M. F., *et al.*, Genetic evidence for archaic admixture in Africa, *Proceedings of the National Academy of Sciences of the United States of America* **108**, 15123–8, 2011; Hsieh, P., *et al.*, Model-based analyses of whole-genome data reveal a complex evolutionary history involving archaic introgression in Central African Pygmies, *Genome Research* **26**, 291–300, 2016.
34 Skoglund, P., *et al.*, Reconstructing prehistoric African population structure, *Cell* **171**, 59–71, 2017.

## 4. LAST HUMAN STANDING

1 Henshilwood, C. S., *et al.*, An abstract drawing from the 73,000-year-old levels at Blombos Cave, South Africa, *Nature* **562**, 115–18, 2018.
2 Brown, K. S., *et al.*, An early and enduring advanced technology originating 71,000 years ago in South Africa, *Nature* **491**, 590–3, 2012.
3 Henshilwood, C. S., *et al.*, A 100,000-year-old ochre-processing workshop at Blombos Cave, South Africa, *Science* **334**, 219–22, 2011.
4 He wasn't really my uncle, and his birth name wasn't Henry. But that's not important right now.
5 Powell, A., *et al.*, Late Pleistocene demography and the appearance of modern human behavior, *Science* **324**, 1298–1301, 2009; D'Errico, F., and Stringer, C. B., Evolution, revolution or saltation scenario for the emergence of modern cultures?, *Philosophical Transactions of the Royal Society of London B* **366**, 1060–9, 2011.
6 Groucutt, H. S., White, T. S., Scerri, E. M. L., *et al.*, Multiple hominin dispersals into Southwest Asia over the past 400,000 years, *Nature* **597**, 376–80, 2021.
7 For a detailed discussion of the human migration into Asia see Dennell, R., *From Arabia to the Pacific: How Our Species Colonised Asia* (Routledge: London and New York, 2020).
8 Akhilesh, K., *et al.*, Early Middle Palaeolithic culture in India around 385–172 ka reframes Out of Africa models, *Nature* **554**, 97–101, 2018.
9 Dennell, R., Human colonization of Asia in the Late

Pleistocene: the history of an invasive species, *Current Anthropology* **58**, Supplement 17, December 2017.
10 Aubert, M., *et al.*, Pleistocene cave art from Sulawesi, Indonesia, *Nature* **514**, 223–7, 2014.
11 Aubert, M., *et al.*, Palaeolithic cave art in Borneo, *Nature* **564**, 254–7, 2018.
12 Kittler, R., *et al.*, Molecular evolution of *Pediculus humanus* and the origin of clothing, *Current Biology* **13**, 1414–17, 2003.

## 5. AGRICULTURE: THE FIRST CASUALTY

1 Botigué, L. R., *et al.*, Ancient European dog genomes reveal continuity since the Early Neolithic, *Nature Communications* **8**, 16082, 2017.
2 Lahtinen, M., *et al.*, Excess protein enabled dog domestication during severe Ice Age winters, *Scientific Reports* **11**, 7, 2021, https://doi.org/10.1038/s41598-020-78214-4.
3 See Diamond, J., Evolution, consequences and future of plant and animal domestication, *Nature* **418**, 700–7, 2002.
4 See Stuart, A. J., *Vanished Giants* (Chicago: University of Chicago Press, 2021).
5 The latest evidence comes from fossilised footprints in New Mexico dated between 23,000 and 20,000 years ago. The date was initially subject to some debate but appears to have been corroborated by later evidence. Other evidence from artefacts in a cave in Mexico suggests an even earlier occupation, as long as 33,000 years ago. See Pigati, J. S., *et al.*, Independent age estimates resolve the controversy of ancient human footprints at White Sands, *Science* **382**, 73–5, 2023; Ardelean, C. F., *et al.*, Evidence of human occupation in Mexico around the Last Glacial Maximum, *Nature* **584**, 87–92, 2020.
6 See Diamond, J., Evolution, consequences and future of plant and animal domestication, *Nature* **418**, 700–7, 2002.
7 Roosevelt, A. C., Population, health, and the evolution of subsistence. In Cohen, M. N., and Armelagos, G. J. (eds), *Paleopathology at the Origins of Agriculture* (Gainesville: University Press of Florida, 2013), pp. 559–83.
8 Alfani, G., and Ó Gráda, C., The timing and causes of famines in Europe, *Nature Sustainability* **1**, 283–8, 2018.

NOTES

9   Unless you have gutted wild animals, it's hard to appreciate just how worm-infested they can be. Those students who have dissected rats and mice will hardly ever encounter this, as those animals are supplied by companies that sell specially bred, disease-free animals to laboratories, and to the pet trade. For example, my pet python (*Python regalis*) eats one mouse every two weeks. Neither the python, nor I, catch wild mice. I buy the mice, frozen, in packs of ten, from a specialist pet store. But I digress. I do not know if animals are dissected at schools today. When I was a pupil, the dogfish we dissected were supplied fresh from the sea (my biology teacher knew the fisherman). But on slicing open the belly of a fresh dogfish (*Scyliorhinus canicula*), the first thing that would appear was a writhing mass of roundworms. It was as if the poor fish was a sports holdall filled with wet spaghetti.

10  Li, J., *et al.*, The emergence, genomic diversity and global spread of SARS-CoV2, *Nature* **600**, 408–18, 2021.

11  Kozlov, M., Huge amounts of bird-flu virus found in raw milk of infected cows, *Nature*, 5 June 2024, https://www.nature.com/articles/d41586-024-01624-1.

12  Various strains of avian influenza have jumped species into humans many times. The pandemic of 1918 – which killed more people than the preceding Great War – originated in a bird host. For the current situation with avian influenza, see https://www.ncbi.nlm.nih.gov/pmc/articles/PMC2095018.

13  Dunne, J., *et al.*, Milk of ruminants in ceramic baby bottles from prehistoric child graves, *Nature* **574**, 246–8, 2019.

14  Evershed, R. P., *et al.*, Dairying, diseases and the evolution of lactose persistence in Europe, *Nature* **608**, 336–45, 2022.

15  Wilkin, S., *et al.*, Dairying pastoralism sustained eastern Eurasian steppe populations for 5,000 years, *Nature Ecology & Evolution* **4**, 346–55, 2020; Wilkin, S., *et al.*, Dairying enabled Early Bronze Age Yamnaya steppe expansions, *Nature* **598**, 629–33, 2021.

16  Diamond, J., The double puzzle of diabetes, *Nature* **423**, 599–602, 2003.

17  This is distinct from type 1, insulin-dependent diabetes.

18  Dowse, G. K., *et al.*, Decline in incidence of epidemic glucose intolerance in Nauruans: implications for the 'thrifty genotype', *American Journal of Epidemiology* **133**, 1093–1104, 1991.

19 Alfani, G., and Ó Gráda, C., The timing and causes of famines in Europe, *Nature Sustainability* **1**, 283–8, 2018.
20 D'Hont, A., *et al.*, The banana (*Musa acuminata*) genome and the evolution of monocotyledonous plants, *Nature* **488**, 213–17, 2012.
21 See https://www.imf.org/en/Blogs/Articles/2022/09/30/global-food-crisis-demands-support-for-people-open-trade-bigger-local-harvests.
22 https://reliefweb.int/report/world/global-food-crisis-what-you-need-know-2023. Source: European Commission's Directorate-General for European Civil Protection and Humanitarian Aid Operations.

## 6. POX-RIDDEN, WORM-EATEN AND LOUSY

1 http:/www.porphyria-patients.uct.ac.za/ppa/types/variegate.
2 https://psychnews.psychiatryonline.org/doi/full/10.1176/pn.40.24.0021.
3 https://www.healthline.com/health/crohns-disease/jewish-ancestry.
4 Fan, J., *et al.*, Gaucher disease protects against tuberculosis, *Proceedings of the National Academy of Sciences of the United States of America* **120**, e2217673120, 2023.
5 Tournebize, R., *et al.*, Reconstructing the history of founder events using genome-wide patterns of allele sharing across individuals, *PLoS Genetics* **18**, e1010243, 2020.
6 Kaessmann, H., *et al.*, Great ape DNA sequences reveal a reduced diversity and an expansion in humans, *Nature Genetics* **27**, 155–6, 2001.
7 Chan, E. K. F., *et al.*, Human origins in a southern African palaeo-wetland and first migrations, *Nature* **575**, 185–9, 2019.
8 Bergström, A., *et al.*, Origins of modern human ancestry, *Nature* **590**, 229–37, 2021.
9 Hu, W., *et al.*, Genomic inference of a severe human bottleneck during the Early to Middle Pleistocene transition, *Science* **381**, 979–84, 2023.
10 Centers for Disease Control, https://www.cdc.gov/prions/vcjd/index.html.
11 https://medlineplus.gov/ency/article/001379.htm.

## NOTES

12 Harper, K., *Plagues Upon the Earth* (Princeton: Princeton University Press, 2021).
13 Irving, A. T., *et al.*, Lessons from the host defences of bats, a unique viral reservoir, *Nature* **589**, 363–70, 2021.
14 Kittler, R., *et al.*, Molecular evolution of *Pediculus humanus* and the origin of clothing, *Current Biology* **13**, 1414–17, 2003.
15 Infections with bacteria, protists and worms can be contained with drugs such as antibiotics, as well as by vaccination. Some drugs exist that can combat viruses, but antibiotics, in general, have no effect on them.
16 Dyer, A., The English sweating sickness of 1551: an epidemic anatomized, *Medical History* **41**, 362–84, 1997.
17 The Norse arrived in North America much earlier – in the eleventh century – but presumably in numbers too small, and in relatively unpopulated areas, to make much impact.
18 https://www.worldometers.info/coronavirus/.
19 Centers for Disease Control, https://www.cdc.gov/ebola/about/index.html.
20 Centers for Disease Control, https://www.cdc.gov/legionella/about.
21 World Health Organization, https://www.who.int/news-room/feature-stories/detail/the-history-of-zika-virus.
22 World Health Organization, https://www.who.int/health-topics/chikungunya#tab=tab_1.
23 European Centre for Disease Prevention and Contol, https://www.ecdc.europa.eu/en/west-nile-fever/facts.
24 I edited the draft text during the UK government's formal inquiry into its handling of the SARS-CoV-19 pandemic, during which evidence was presented suggesting that the government was ill-prepared for the pandemic and might also be at a loss should another pandemic appear.
25 At this point in the draft text, my friend Brian Clegg wrote, 'Not a cheerful read, is it?' The even less cheerful paragraph that follows was one I added later.
26 Karlsson, E. K., *et al.*, Natural selection and infectious disease in human populations, *Nature Reviews Genetics* **15**, 379–93, 2014.

NOTES

### 7. ON THE BRINK

1. Burger, J. R., and Fristoe, T. S., Hunter-gatherer populations inform modern ecology, *Proceedings of the National Academy of Sciences of the United States of America* **115**, 1137–9, 2018.
2. Ehrlich, P. R., *The Population Bomb* (New York: Ballantine Books, 1968).
3. Ibid., p. 92.
4. Ibid., pp. 135–6.
5. Ibid., p. 165.
6. It might be no coincidence that many of these polities were, in effect, creations imposed on the native populations by colonial powers insensitive to the cultural and ethnic divisions within them.
7. Ibid., p. 139.
8. Ibid., p. 198.
9. https://www.smithsonianmag.com/innovation/book-incited-worldwide-fear-overpopulation-180967499/.
10. Ibid.
11. Friedman, J., *et al.*, Measuring and forecasting progress towards the education-related SDG targets, *Nature* **580**, 636–9, 2020.
12. It is perhaps telling, if only of the evanescence of nations, that neither Czechoslovakia nor the Soviet Union still exists as a polity.
13. https://www.macrotrends.net/countries/WLD/world/population-growth-rate.
14. I'm joking about rampaging artificially intelligent killer robots, and the invasion of malevolent aliens. Though one never knows.
15. The source of the aphorism is unclear. Many ascribe it to baseball player Yogi Berra, but I have seen it attributed to film-maker Woody Allen and physicist Niels Bohr.
16. The predictions of Murray and colleagues can be found in Vollset, S. E., *et al.*, Fertility, mortality, migration and population scenarios for 195 countries and territories from 2017 to 2100: a forecasting analysis for the Global Burden of Disease Study, *The Lancet* **396**, 1285–1306, 2020.
17. The goals include universal secondary education and universal availability of contraceptives by 2030.

18 Murray and colleagues, however, prefer another measure, CCF50. This is the average number of children born to a woman in a given birth cohort if she lives to the end of her reproductive lifespan. This tends to be a stabler measure than TFR.
19 https://www.nytimes.com/2023/01/16/business/china-birth-rate.html.
20 Murray and colleagues define working age as between twenty and sixty-four years old.
21 Basten, S., *et al.*, Very long range global population scenarios to 2300 and the implications of sustained low fertility, *Demographic Research* **28**, 1145–66, 2013.
22 https://www.icaew.com/insights/viewpoints-on-the-news/2022/nov-2022/business-confidence-weakest-since-pandemic-finds-icaew.
23 https://www.imf.org/en/Blogs/Articles/2022/10/11/policymakers-need-steady-hand-as-storm-clouds-gather-over-global-economy.
24 https://www.worldbank.org/en/news/video/2023/01/10/global-economic-prospects-slowdown-growth-risks-economy-expert-answers.
25 https://www.weforum.org/agenda/2016/07/millennials-will-be-the-first-generation-to-earn-less-than-their-parents/.
26 A recent case was an advertisement for someone to share not an apartment in the Canadian city of Toronto, or a room, but a bed – for 900 Canadian dollars a month. The advertisement went viral, and highlighted the acute shortage of rental properties in major cities. https://economictimes.indiatimes.com/news/international/us/toronto-rent-ca900-for-half-bed-watch-viral-video-know-everything-about-it/articleshow/105449273.cms?from=mdr.
27 Masood, E., *GDP: The World's Most Powerful Formula and Why It Must Now Change* (London: Icon Books, 2021).
28 Carlsen, E., *et al.*, Evidence for decreasing quality of semen during past 50 years, *British Medical Journal* **305**, 609–13, 1992.
29 Levine, H., *et al.*, Temporal trends in sperm count: a systemic review and meta-regression analysis of samples collected globally in the 20th and 21st centuries, *Human Reproduction Update* **29**, 157–76, 2022.
30 Levine, H., *et al.*, Temporal trends in sperm count: a systemic

review and meta-regression analysis, *Human Reproduction Update* **23**, 646–59, 2017.
31 Lv, M.-Q., *et al.*, Temporal trends in semen concentration and count among 327,323 Chinese healthy men from 1981 to 2019, *Human Reproduction* **36**, 1751–75, 2021.
32 Skakkebæk, N. E., *et al.*, Environmental factors in declining human fertility, *Nature Reviews Endocrinology* **18**, 139–57, 2022.
33 Ajayi, A. B., *et al.*, Low sperm counts: biophysical profiles of oligospermic males in sub-Saharan Africa, *Open Journal of Urology* **8**, 228–47, 2018.
34 Sengupta, P., *et al.*, Evidence for decreasing sperm count in African population from 1965 to 2015, *African Health Sciences* **17**, 418–27, 2017.
35 https://www.un.org/uk/desa/68-world-population-projected-live-urban-areas-2050-says-un.
36 Ilacqua, A., *et al.*, Lifestyle and fertility: the influence of stress and quality of life on male fertility, *Reproductive Biology and Endocrinology* **16**, 115, 2018.

## 8. OVER THE EDGE

1 https://en.wikipedia.org/wiki/List_of_lost_settlements_in_Norfolk.
2 https://www.bgs.ac.uk/case-studies/coastal-erosion-at-happisburgh-norfolk-landslide-case-study/
3 https://www.bbc.co.uk/news/uk-england-norfolk-63822899#.
4 http://happisburgh.org.uk/ccag-press/plan-to-allow-sea-to-flood-norfolk-villages/
5 Shennan, I., *et al.*, Late Holocene relative land- and sea-level changes: providing information for stakeholders, *GSA Today* **19**, 52–3, 2009.
6 https://www.edp24.co.uk/lifestyle/21095516.floods-1953-graphic—unfolded/#.
7 https://www.theccc.org.uk/2018/10/26/current-approach-to-protecting-englands-coastal-communities-from-flooding-and-erosion-not-fit-for-purpose-as-the-climate-changes/.
8 Vince, G., *Nomad Century: How to Survive the Climate Upheaval* (London: Allen Lane, 2022).

## NOTES

9 https://www.climatechangepost.com/news/multi-century-sea-level-rise-may-lead-to-unprecedented-threats-to-coastal-cities.
10 https://qz.com/1700769/sea-level-rise-is-set-to-flood-un-headquarters-as-soon-as-2100.
11 https://thestarryeye.typepad.com/weather/2012/10/hurricanes-tropical-storms-that-have-impacted-new-york-city-1979–2011.html.
12 https://www.americanprogress.org/article/when-you-cant-go-home/.
13 https://www.maplecroft.com/insights/analysis/asian-cities-in-eye-of-environmental-storm-global-ranking/.
14 https://www.pbs.org/newshour/world/why-indonesia-is-moving-its-capital-from-jakarta-to-borneo.
15 https://www.metoffice.gov.uk/weather/learn-about/weather/case-studies/heatwave.
16 https://www.isws.illinois.edu/statecli/general/1995chicago.htm.
17 https://www.bbc.co.uk/news/world-asia-65518528.
18 https://edition.cnn.com/2024/06/02/india/india-heatwave-poll-worker-deaths-intl-hnk/index.html.
19 https://www.bbc.co.uk/news/science-environment-65403381.
20 https://climate.copernicus.eu/copernicus-2023-hottest-year-record.
21 Sherwood, S. C., et al., An adaptability limit to climate change due to heat stress, *Proceedings of the National Academy of Sciences of the United States of America* **107**, 9552–5, 2010.
22 Pal, J. S., and Eltahir, E. A. B., Future temperature in southwest Asia projected to exceed a threshold for human adaptability, *Nature Climate Change* **6**, 197–200, 2016.
23 https://amp.theguardian.com/world/article/2024/jun/18/hundreds-of-hajj-pilgrims-die-in-mecca-from-heat-related-illness.
24 Pal, J. S., and Eltahir, E. A. B., Future temperature in southwest Asia projected to exceed a threshold for human adaptability, *Nature Climate Change* **6**, 197–200, 2016.
25 Kang, S., and Eltahir, E. A. B., North China Plain threatened by deadly heatwaves due to climate change and irrigation, *Nature Communications* **9**, 2894, 2018.
26 Im, E.-S., et al., Deadly heat waves projected in densely populated agricultural regions of South Asia, *Science Advances* **3**, e1603322, 2017.

NOTES

27 Mora, C., *et al.*, Global risk of deadly heat, *Nature Climate Change* **7**, 501–6, 2017.
28 Coffel, E. D., *et al.*, Temperature and humidity based projections of a rapid rise in global heat stress exposure during the 21st century, *Environmental Research Letters* **13**, 014001, 2018.
29 Hsiang, S. M., *et al.*, Quantifying the influence of climate on human conflict, *Science* **341**, 2013, https://doi.org/10.1126/science.1235367.
30 Vince, G., *Nomad Century: How to Survive the Climate Upheaval* (London: Allen Lane, 2022).
31 O'Neill, D. W., *et al.*, A good life for all within planetary boundaries, *Nature Sustainability* **1**, 88–95, 2018.

## 9. FREE FALL, AND AFTER

1 Basten, S., *et al.*, Very long range global population scenarios to 2300 and the implications of sustained low fertility, *Demographic Research* **28**, 1145–66, 2013.
2 https://populationconnection.org/blog/world-population-milestones-throughout-history/.
3 Gott, J. R., Implications of the Copernican principle for our future prospects, *Nature* **363**, 315–19, 1993; discussed in a 2017 article in *The Washington Post* by Christopher Ingraham, https://www.washingtonpost.com/news/wonk/wp/2017/10/06/we-have-a-pretty-good-idea-of-when-humans-will-go-extinct/.
4 At the time of writing (1 December 2023), NASA lists 4,135 planetary systems beyond ours, which contain 5,550 confirmed planets, with another 10,009 possible planetary sightings awaiting confirmation. However, the Earth is unique (so far) in being the only planet with life on it. There is no intelligent life on Earth, however, as my son says that he is only here until the Lizard People return to claim him for their own. https://exoplanets.nasa.gov.
5 For younger readers, the Berlin Wall was a Cold War–era edifice that once divided the city of Berlin in Germany into western and eastern sectors.
6 Why 95 per cent? There is no special reason, except that the convention in statistics is to regard a result as being significant if the odds of it happening by chance are 5 per cent (1 in 20)

or less (that's 100 per cent minus 95 per cent). A significant result, therefore, is privileged, or special. Conversely, there is nothing special about an event with odds of 95 per cent (19 in 20). Such a happening is a virtual certainty – such as turning up at the Berlin Wall at a point in its existence that was not especially close to its beginning or to its end. The use of 95 per cent is purely conventional. Depending on what one wishes to find out, tests of significance can be set at 99 per cent (1 in 100), or 99.9 per cent (1 in 1,000), or any value you please. I should add that significance tests like this are falling out of fashion, in favour of other, more refined statistical methods, such as Bayesian statistics. For an in-depth discussion of Bayesian statistics that's accessible to the layperson, see *Everything Is Predictable* by Tom Chivers (New York: Simon & Schuster, 2024).

7   Gott's figures were 5,128 and 7,800,000 years, but that was based on the appearance of *Homo sapiens* around 200,000 years ago, which was the best that was known in 1993.

8   Footnote for Bayes fans: Given what we know about the tenure of other species of hominins on Earth, this means that we should probably adjust our 'prior' (short for 'prior probability') from twelve million years to something much less. In other words, given that most hominin species seem to exist for no more than a million to two million years, it seems a fair starting assumption that *Homo sapiens* will last for a similar length of time. On the other hand, *Homo sapiens* has achieved far more, in a much shorter frame of time, than any other hominin species we know about, which calls this assumption into question. That's the trouble with priors.

9   Arsuaga, J.-L., *et al.*, Three new human skulls from the Sima de los Huesos Middle Pleistocene site in Sierra de Atapuerca, Spain, *Nature* **362**, 534–7, 1993.

10  Higham, T., *et al.*, The timing and spatiotemporal patterning of Neanderthal disappearance, *Nature* **512**, 306–9, 2014.

11  Rizal, Y., *et al.*, Last appearance of *Homo erectus* at Ngandong, Java, 117,000–108,000 years ago, *Nature* **577**, 381–5, 2020.

12  Tilman, D., *et al.*, Habitat destruction and the extinction debt, *Nature* **371**, 65–6, 1994.

13  Krausmann, F., *et al.*, Global human appropriation of net primary production doubled in the 20th century, *Proceedings*

## NOTES

*of the National Academy of Sciences of the United States of America* **110**, 10324–9, 2013. I'll discuss photosynthesis and its importance to human survival later in this book.

14  Bar-On, Y. M., *et al.*, The biomass distribution on Earth, *Proceedings of the National Academy of Sciences of the United States of America* **115**, 6506–11, 2018.

15  Before I go any further, I should note that the story of the pre-contact collapse of the society and ecology on Rapa Nui has lately been questioned. DiNapoli, R. J., *et al.*, A model-based approach to the tempo of "collapse": the case of Rapa Nui (Easter Island), *Journal of Archaeological Science* **116**, 105094, 2020.

16  Gee, H., Treeless at Easter, *Nature* **431**, 411, 2004.

17  Diamond, J., Twilight at Easter, *The New York Review of Books*, 25 March 2004.

18  Higham, T., *et al.*, The timing and spatiotemporal patterning of Neanderthal disappearance, *Nature* **512**, 306–9, 2014.

19  See Svante Pääbo's book *Neanderthal Man: In Search of Lost Genomes* (New York: Basic Books, 2015) for a lively and accessible account of this research.

20  Vernot, B., *et al.*, Unearthing Neanderthal population history using nuclear and mitochondrial DNA from cave sediments, *Science* **372**, 2021, https://doi.org/10.1126/science.abf1667

21  Hajdinjak, M., *et al.*, Reconstructing the genetic history of late Neanderthals, *Nature* **555**, 652–6, 2018.

22  Harris, K., and Nielsen, R., The genetic cost of Neanderthal introgression, *Genetics* **203**, 881–91, 2016.

23  Rios, L., *et al.*, Possible further evidence of low genetic diversity in the El Sidrón (Asturias, Spain) Neandertal group: congenital clefts of the atlas, *PLoS ONE* **10**(9), e0136550, 2015.

24  Prüfer, K., *et al.*, The complete genome sequence of a Neanderthal from the Altai mountains, *Nature* **505**, 43–9, 2014.

25  Skov, L., *et al.*, Genetic insights into the social organization of Neanderthals, *Nature* **610**, 519–25, 2022.

26  Vaesen, K., *et al.*, Inbreeding, Allee effects and stochasticity might be sufficient to account for Neanderthal extinction, *PLoS ONE* **14**(11), e0225117, 2019; Kolodny, O., and Feldman, M. W., A parsimonious neutral model suggests Neanderthal replacement was determined by migration and random species drift, *Nature Communications* **8**, 1040, 2017.

## 10. THE FUTURE IS GREEN AND FEMALE

1 It's worth saying, however, that Gott said quite plainly that humanity was extremely unlikely to become a spacefaring species to any significant degree.
2 Astronauts staying in space for extended periods without the sensation of weight experience a host of unwelcome physiological conditions such as bone loss, muscle wastage and cardiovascular problems. The record at the time of writing is 371 days. https://www.bbc.com/future/article/20230927-what-a-long-term-mission-in-space-does-to-the-human-body.
3 Ditto faster-than-light travel, indefinite cryonic suspension or hibernation of humans, generation starships and so on.
4 Brannen, P., *The Ends of The World* (London: Oneworld, 2017).
5 Benton, M. J., *When Life Nearly Died* (London: Thames and Hudson, 2003).
6 Hannah, M., *Extinctions: Living and Dying in the Margin of Error* (Cambridge: Cambridge University Press, 2021).
7 Barnosky, A., *et al.*, Has the Earth's sixth mass extinction already arrived?, *Nature* **471**, 51–7, 2011.
8 https://www.pbs.org/wgbh/evolution/library/03/2/l_032_04.html.
9 Tallavaara, M., *et al.*, Productivity, biodiversity and pathogens influence the global hunter-gatherer population density, *Proceedings of the National Academy of Sciences of the United States of America* **115**, 1232–7, 2018.
10 Burger, J. R., and Fristoe, T. S., Hunter-gatherer populations inform modern ecology, *Proceedings of the National Academy of Sciences of the United States of America* **115**, 1137–9, 2018.
11 At the time, neither Ehrlich nor anyone else would have known that the rate of population growth might already be slowing.
12 Ehrlich, P., *The Population Bomb* (New York: Ballantine Books, 1968), pp. 105–6.
13 Evenson, R. E., and Gollin, D., Assessing the impact of the Green Revolution, 1960 to 2000, *Science* **300**, 758–62, 2003; Pingali, P. L., Green Revolution: impacts, limits and the path ahead, *Proceedings of the National Academy of Sciences of the United States*

*of America* **109**, 12302–8, 2012; Khush, G. S., Green Revolution, the way forward, *Nature Reviews Genetics* **2**, 815–22, 2001.

14 Cohen, J. E., How many people can the Earth support?, *The Sciences* **35**(5), 2–49, 1995.

15 Ibid., p. 19.

16 Rockström, J., *et al.*, A safe operating space for humanity, *Nature* **461**, 472–5, 2009. The researchers updated and refined their analysis in 2015 (Steffen, W., *et al.*, Planetary boundaries: guiding human development on a changing planet, *Science* **347**, 1259855, 2015).

17 Almost four fifths of the atmosphere consists of the gas nitrogen, which normally does not react with anything. However, it can be converted into a form that's available for use by living things in two ways. First, in the roots of leguminous plants such as peas and beans; and second, by an industrial process known as the Haber–Bosch process, in which nitrogen from the air is reacted with hydrogen to create ammonia.

18 Rockström, J., *et al.*, Safe and just Earth system boundaries, *Nature* **619**, 102–11, 2023, https://doi.org/10.1038/s41586-023-06083-8.

19 Costanza, R., *et al.*, The value of the world's ecosystem services and natural capital, *Nature* **387**, 253–60, 1997.

20 Biosphere 2 is an environmental research facility currently run by the University of Arizona. It was originally conceived as a sealed, self-sustaining biosphere designed to test ideas for self-sustaining life in space colonies. People lived inside Biosphere 2 for varying periods of time with, it has to be said, mixed success. I shall discuss Biosphere 2 more in the final chapter. https://biosphere2.org/about/about-biosphere-2.

21 Costanza, R., *et al.*, Twenty years of ecosystem services: how far have we come and how far do we still need to go?, *Ecosystem Services* **28**, 1–16, 2017.

22 I shall discuss some of these in the next chapter.

23 Abel, G. J., *et al.*, Meeting the Sustainable Development Goals leads to lower world population growth, *Proceedings of the National Academy of Sciences of the United States of America* **113**, 14294–9, 2016.

24 Friedman, J., *et al.*, Measuring and forecasting progress towards the education-related SDG targets, *Nature* **580**,

636–9, 2020; Grant, M., Monitoring global education inequality, *Nature* **580**, 591–2, 2020.
25  Friedman, J., et al., Measuring and forecasting progress towards the education-related SDG targets, *Nature* **580**, 636–9, 2020.
26  Liao, P. V., and Dollin, J., Half a century of the oral contraceptive pill, *Canadian Family Physician* **58**, e757–60, 2012.

## 11. TURNING OVER A NEW LEAF

1  Krausmann, F., et al., Global human appropriation of net primary production doubled in the 20th century, *Proceedings of the National Academy of Sciences of the United States of America* **110**, 10324–9, 2013.

2  Material for the discussion on how photosynthesis might be improved comes from Zhu, X.-G., et al., Improving photosynthetic efficiency for greater yield, *Annual Reviews of Plant Biology* **61**, 235–61, 2010; Blankenship, R. E., et al., Comparing photosynthetic and photovoltaic efficiencies and recognizing potential for improvement, *Science* **332**, 805–9, 2011; Ort, D. R., et al., Redesigning photosynthesis to sustainably meet global food and bioenergy demand, *Proceedings of the National Academy of Sciences of the United States of America* **112**, 8529–36, 2015; Long, S. P., et al., Meeting the global food demand of the future by engineering crop photosynthesis and yield potential, *Cell* **161**, 56–66, 2015; Betti, M., et al., Manipulating photorespiration to increase plant productivity: recent advances and perspectives for crop improvement, *Journal of Experimental Botany* **67**, 2977–88, 2016; Sharwood, R. E., Engineering chloroplasts to improve Rubisco catalysis: prospects for translating improvements into food and fiber crops, *New Phytologist* **213**, 494–510, 2017; Bailey-Serres, J., et al., Genetic strategies for improving crop yields, *Nature* **575**, 109–18, 2019; and Kolbert, E., Creating a better leaf, *The New Yorker*, 13 December 2021.

3  Figures from Blankenship, R. E., et al., Comparing photosynthetic and photovoltaic efficiencies and recognizing potential for improvement, *Science* **332**, 805–9, 2011.

4  For a useful primer on photosynthesis, see Johnson, M. P., Photosynthesis, *Essays in Biochemistry* **60**, 255–73, 2016.

# NOTES

5   Calculating efficiency is a rather fraught process – it depends on how you define it. The figure of 1 per cent comes from comparing the energy harvested annually from the Sun for a given amount of biomass with the heat required to burn the products of photosynthesis in that biomass (glucose) to carbon dioxide and water under certain standard conditions. Efficiencies measured during the growing season are higher – up to about 4.3 per cent – but still fall far short of the theoretical maximum of 12 per cent. Figures from Blankenship, R. E., et al., Comparing photosynthetic and photovoltaic efficiencies and recognizing potential for improvement, *Science* **332**, 805–9, 2011.

6   Blankenship, R. E., Early evolution of photosynthesis, *Plant Physiology* **154**, 434–8, 2010.

7   You might have seen these referred to in older texts as 'blue-green algae'. However, they are bacteria, and much simpler in structure than algae.

8   There are also a lot of eukaryotes that are microscopic. These include familiar pond life such as the amoebas and paramecia ('slipper animalcules') as well as less familiar (but important) residents of the sea, such as dinoflagellates and diatoms, and some notorious disease-causing creatures such as the malaria parasite.

9   In this case of spinach, though the figure will be similar for other green plants.

10  Although hardly a major crop, it turns out that tobacco plants are relatively easy to manipulate by genetic engineering, at least when compared with rice, corn or wheat.

11  See, for example, Haas, T., et al., Technical photosynthesis involving $CO_2$ electrolysis and fermentation, *Nature Catalysis* **1**, 32–9, 2018.

12  See, for example, Liu, C., et al., Water splitting–biosynthetic system with $CO_2$ reduction efficiencies exceeding photosynthesis, *Science* **352**, 1210–13, 2016.

13  Cai, T., et al., Cell-free chemoenzymatic starch synthesis from carbon dioxide, *Science* **373**, 1523–7, 2021.

14  Romero Cuellar, N. S., et al., Two-step electrochemical reduction of $CO_2$ towards multi-carbon products at high current densities, *Journal of $CO_2$ Utilization* **36**, 263–75, 2020.

15 Hann, E. C., *et al.*, A hybrid inorganic–biological artificial photosynthesis system for energy-efficient food production, *Nature Food* **3**, 461–71, 2022; Wang, T., and Gong, J., Artificial photosynthesis of food from $CO_2$, *Nature Food* **3**, 409–10, 2022.
16 Shepon, A., *et al.*, The opportunity cost of animal based diets exceeds all food losses, *Proceedings of the National Academy of Sciences of the United States of America* **115**, 3804–9, 2018.
17 Jones, N., The new proteins coming to your plate, *Nature* **619**, 26–8, 2023.
18 Disclaimer – I am not a vegetarian. Though I do have poultry in my backyard that produce fresh eggs, reducing supply-chain losses in the Gee household egg budget to zero.

## 12. EXPANDING THE HUMAN NICHE

1 Erickson, C. L., An artificial landscape fishery in the Bolivian Amazon, *Nature* **408**, 190–3, 2000.
2 Prümers, H., *et al.*, Lidar reveals pre-Hispanic low-density urbanism in the Bolivian Amazon, *Nature* **606**, 325–8, 2022.
3 https://www.forestresearch.gov.uk/tools-and-resources/statistics/forestry-statistics/forestry-statistics-2018/woodland-areas-and-planting/woodland-area-2/area-of-woodland-changes-over-time/.
4 Song, X.-P., *et al.*, Global land change from 1982 to 2016, *Nature* **560**, 639–43, 2018.
5 Boivin, N. L., *et al.*, Ecological consequences of human niche construction: examining long-term anthropogenic shaping of global species distributions, *Proceedings of the National Academy of Sciences of the United States of America* **113**, 6388–96, 2016.
6 Benton, T., Oceans of garbage, *Nature* **352**, 113, 1991.
7 Cook, J. M., and West, S. A., Figs and fig wasps, *Current Biology* **15**, R978–980, 2005.
8 For a review, see, for example, Zimov, S. A., *et al.*, Mammoth steppe, a high-productivity phenomenon, *Quaternary Science Reviews* **57**, 26–45, 2012.
9 Stuart, A. J., *et al.*, Pleistocene to Holocene extinction dynamics in giant deer and woolly mammoth, *Nature* **431**, 684–9, 2004.
10 Boivin, N. L., *et al.*, Ecological consequences of human niche construction: examining long-term anthropogenic shaping of

global species distributions, *Proceedings of the National Academy of Sciences of the United States of America* **113**, 6388–96, 2016.
11 Ibid.
12 As in the song 'Big Yellow Taxi' by Joni Mitchell.
13 Thomas, C. D., *Inheritors of the Earth: how nature is thriving in an age of extinction* (New York: PublicAffairs, 2017).
14 Xu, C., *et al.*, Future of the human climate niche, *Proceedings of the National Academy of Sciences of the United States of America* **117**, 11350–5, 2020.
15 Vince, G., *Nomad Century: How to Survive the Climate Upheaval* (London: Allen Lane, 2022).
16 Zink, K. D., and Lieberman, D. E., Impact of meat and Lower Palaeolithic food processing techniques on chewing in humans, *Nature* **531**, 500–3, 2016.
17 See, for example, Schmidt, P., *et al.*, Production method of the Königsaue birch tar documents cumulative culture in Neanderthals, *Archaeological and Anthropological Sciences* **15**, 84, 2023.
18 See, for example, Thompson, J. C., *et al.*, Early human impacts and ecosystem reorganization in southern-Central Africa, *Science Advances* **7**, 2021, https://doi.org/10.1126/sciadv.abf9776.
19 Kittler, R., *et al.*, Molecular evolution of *Pediculus humanus* and the origin of clothing, *Current Biology* **13**, 1414–17, 2003.
20 Around 56 per cent of the population (4.4 billion people) now live in cities, and the proportion is expected to rise. Seven in ten people will be city dwellers in 2050. https://www.worldbank.org/en/topic/urbandevelopment/overview.
21 Egregious exceptions to this rule are the social insects – wasps, bees, ants and termites – that live in gigantic nests or colonies every bit as artificial, in their way, as human cities.
22 Chappell, S. P., *et al.*, NEEMO 15: evaluation of human exploration systems for near-Earth asteroids, *Acta Astronautica* **89**, 166–78, 2013.
23 Cornelius, K., Biosphere 2: the once infamous live-in terrarium is transforming climate research, *Scientific American*, 4 October 2021; Zimmer, K., The lost history of one of the world's strangest science experiments, *The New York Times*, 29 March 2019. I do not know whether the fact that the acreage of Biosphere 2 is almost equivalent to pi is significant.
24 Anyone who has ever tried to maintain an aquarium in the

home knows that small ones are prone to instability. At home I have a fifty-gallon freshwater tropical tank containing several angelfish and one large, very old catfish. Because these fish are tropical, the temperature is maintained by a heater and thermostat which, once set, can be forgotten about. The volume is so large that the tank requires almost no maintenance. Matters were different when I had a small cold-water tank on my desk that contained a single goldfish. This tank contained less than five gallons. It's much harder to keep a tank cool than warm, and this is especially true when the volume is small. Without constant vigilance the tank soon became clogged with algae, which would have used up all the dissolved oxygen and suffocated the fish. I had to change the water completely at least once a week. The larger tank can go for months with only partial water changes.

25 See Chen, M., *et al.*, Review of space habitat designs for long term space explorations, *Progress in Aerospace Sciences* **122**, 100692, 2021.

26 https://www.nasa.gov/specials/artemis/.

27 https://www.unoosa.org/oosa/en/ourwork/spacelaw/treaties/introouterspacetreaty.html.

28 See Tim Marshall's book *The Future of Geography* (London: Elliott and Thompson, 2023) for a look at the near future of astropolitics. A highly readable but in-depth treatment of the legal obstacles that will beset efforts to settle space can be found in *A City on Mars: Can We Settle Space, Should We Settle Space, and Have We Really Thought This Through?* by Kelly and Zach Weinersmith (London: Particular Books, 2023).

29 Six technologies NASA is advancing to send humans to Mars, https://www.nasa.gov/directorates/spacetech/6_Technologies_NASA_is_Advancing_to_Send_Humans_to_Mars.

30 See Chen, M., *et al.*, Review of space habitat designs for long term space explorations, *Progress in Aerospace Sciences* **122**, 100692, 2021.

31 https://solarsystem.nasa.gov/asteroids-comets-and-meteors/asteroids/overview/?page=0&per_page=40&order=name+asc&search=&condition_1=101%3Aparent_id&condition_2=asteroid%3Abody_type%3Ailike.

32 See, for example, Chapman, C. R., and Morrison, D., Impacts

on the Earth by asteroids and comets: assessing the hazard, *Nature* **367**, 33–40, 1994. In the article, the authors estimate that the chance of an asteroid of 1.24 miles or more in diameter striking the Earth in the next century – an impact large enough to cause severe ecological damage and kill a large fraction of the Earth's human population – is about one in 10,000. The chance is remote, but not impossible. It's about the same as the chance of being struck by lightning during one's life (one in 15,300) or finding a four-leaf clover (1 in 10,000). The chances of dying in an airplane crash are about one in eleven million. https://stacker.com/art-culture/odds-50-random-events-happening-you.

33 Rao, R., Smashing success: humanity has diverted an asteroid for the first time, *Nature*, 11 October 2022, https://www.nature.com/articles/d41586-022-03255-w.

34 It's been a standing joke among science journalists for decades that nuclear fusion will join the existing mix of energy sources in any routine way in another thirty years from the time of writing. That having been said, experiments on nuclear fusion in 2022, for the first time, created a reaction that made more energy than was put in to drive it. See Tollefson, J., and Gibney, E., Nuclear-fusion lab achieves 'ignition': what does it mean?, *Nature* **612**, 597–8, 2022.

35 Ross, S. D., Near-Earth asteroid mining, Space industry report, https://citeseerx.ist.psu.edu/document?repid=rep1&type=pdf&doi=8c342ea48e59acd5649a7758add94a6e025d7a7e.

36 Grandl, W., and Böck, C., Asteroid habitats – living inside a hollow celestial body. In: Badescu, V., Zacny, K., Bar-Cohen, Y. (eds.), *Handbook of Space Resources* (Springer, 2023), https://doi.org/10.1007/978-3-030-97913-3_22.

37 https://worldpopulationreview.com/boroughs/manhattan-population.

38 Miklavcic, P. M., *et al.*, Habitat Bennu: design concepts for spinning habitats constructed from rubble pile near-Earth asteroids, *Frontiers in Astronomy and Space Sciences* **8**, 645363, 2022.

39 Weinersmith, K., and Weinersmith, Z., *A City on Mars: Can We Settle Space, Should We Settle Space, and Have We Really Thought This Through?* (London: Particular Books, 2023).

## NOTES

40  This is attributed to George Mallory, who attempted to scale Everest in 1924. It is not known whether he succeeded – he perished during the expedition. This should really be a cautionary tale for explorers in space. https://www.alamosa.org/travel-tools-tips/a-travelers-blog/700-because-it-is-there.

41  Although England and Scotland were ruled by the same monarch, James I (James VI of Scotland) after 1603, the countries and their governments remained entirely separate for another century.

42  Gott, J. R., Implications of the Copernican principle for our future prospects, *Nature* **363**, 315–19, 1993; discussed in a 2017 article in the *Washington Post* by Christopher Ingraham, https://www.washingtonpost.com/news/wonk/wp/2017/10/06/we-have-a-pretty-good-idea-of-when-humans-will-go-extinct/.

### AFTERWORD

1  Gee, H., Humans are doomed to go extinct, *Scientific American*, 21 November 2021, https://www.scientificamerican.com/article/humans-are-doomed-to-go-extinct/.

# Index

*2001: A Space Odyssey* (Clarke), 61, 212–213, 222

*A (Very) Short History of Life on Earth* (Gee), 7, 104, 234
abortion, 117, 128
Africa: cities in, 143; Cradle of Humankind, 44; genetic diversity in, 50–51, 67–68, 104; human migration from, 19; human origin in, 56, 60–61; sperm quality in, 134–135; sub-Saharan, 123–125, 131, 180; West, 69, 89, 108, 148
Afrikaners, 16, 100, 101
agriculture: health and, 15–17, 71, 91–97, 108–109; human evolution/population and, 93–94, 188–190; invention, development of, 71, 88–89, 120, 149, 150, 173, 176–177, 210–211
air conditioning, 213–214
Alexander, Robert McNeill, 31, 242n22
Altamira, 78
Amazon, 203, 204, 206
Amish, 16, 101–102
animal product consumption, 201
Antarctic Treaty, 228–229
apes, 24, 34–37, 40; *see also* specific type
Aquarius (submarine habitat), 216–217, 219
Aquatic Ape hypothesis, 241n20
*Ardipithecus kadabba, ramidus*, 37
*Artemis* (space program), 220, 221–222
Ashkenazi Jews, 16, 102–103
asteroids, 172, 223–226
*Australopithecus*, genus, 45, 46, 55–56
*Australopithecus afarensis*, 32–33, 37, 41

Bab el Mandeb, strait of, 75
backbone, 10, 36, 38–39, 41
bacteria, 106, 107, 108
Bakker, Robert, 4
Ball, Philip, 246
bananas, 15, 97
barbary ape, 24

bats, 109
Bayesian statistics, 259n6
Beard, Mary, 238n11
beef production, 201
Berlin Wall, 155–156
biodiversity, 208–209
Biosphere 2, 183, 217–218, 219
bipedality, 10, 33, 36–42, 238n14
bipolar disorder, 16, 101–102
birds, 39, 109
Black Death, 96, 112, 150
Blombos cave, 72–73, 78
body louse, 79, 109, 211
Boivin, Nicole, 208, 209
Borneo cave art, 78
bovine spongiform encephalopathy (BSE), 107
brain power, 70, 151, 174
*A Brief History of Everyone Who Has Ever Lived* (Rutherford), 246
Brown, Louise, 134

C3, C4 plants, 198
carbon dioxide, 197–198, 207
Carboniferous Period, 4
Carlsen, Elizabeth, 132–133
carrying capacity, 89, 114, 176–177, 178, 181–182, 184
Carson, Rachel, 117
Caucasus mountains, 47
Charles II ('El Hechizado'), King of Spain, 53–55
chimpanzees, 16, 107–108
China, People's Republic of: agriculture, 188–189; Great Leap Forward, 16, 96, 123; population, 118, 123, 128, 133, 134; weather, 147

Chivers, Tom, 259n6
chlorophyll, 193–194, 196, 197
chloroplasts, 192–193
*Chororapithecus*, 35
*A Christmas Carol* (Dickens), 171
cities: as foci of disease, 108; as prototypes for spacecraft, 215, 218, 225; climate change, 141–143, 151, 211, 213–214, 215; countryside vs., 204; housing, 132; population in 113, 135
*A City on Mars* (Weinersmith & Weinersmith), 226, 268n28
Clarke, Arthur C., 61
Clegg, Brian, 85, 254n25
climate change: Climate Change Committee (UK), 140; climate justice, 182; driver of migration, 135, 137–138, 149–150; extreme weather events, 139–141, 201, 209–210; influence of agriculture, 89; source of conflict, threat, 18–19, 25, 67, 148, 152, 167, 213–214; unpredictability of, 140
clothing invention, 79
Cohen, Joel E., 181
Conan Doyle, Sir Arthur, 47
contraception, 118, 122, 129, 186–187
Cook, Captain James, 161
Copernicus, Nicolaus, 155
Costanza, Robert, 183–184
COVID-19 (SARS-CoV-19), 16, 93, 109, 112, 130
Cradle of Humankind, 44

Cretaceous Period, 1, 2, 175
Cretan footprints, 33–34
Creutzfeldt–Jakob disease, 107
Crohn's disease, 16, 102–103
Cromer, 87, 137, 139
cyanobacteria, 192, 193

*Danuvius*, 34, 40
Darwin, Charles, 55, 239n32
demographic transition, 151
Denisovans, 8, 14, 49, 51–52, 64, 65–67, 69, 80–82, 237n9
Dennell, Robin, 74–75, 77–78
depression, 53–55
DeSilva, Jeremy, 238n14
Devonian Period, 4
Diamond, Jared, 21, 91, 94–95, 161–162
dinosaurs, 1–5, 32, 39, 175, 223, 224
diabetes (type 2), 94–95
diseases, 16, 105–109, 112–113; *see also* specific disease
Dickens, Charles, 171
Domesday Book, 137
domestication, animals and plants, 14–15, 86–87, 89, 91, 177
DNA (deoxyribonucleic acid), 51, 52, 58–60, 64–65, 237n9
Dutch East India Company, 99

Earth system processes, 182
Easter Island, 160–162
Ebu Gogo, 50
economy, 130–131, 132, 151–152
ecosystem services, 182–184
education access, 185–186

Ehrlich, Paul, 17, 22–23, 115–120, 136, 153, 173, 177–178, 181
English sweating sickness, 111
enzymes, 193, 194
erosion, 137–140
*Essay on the Principle of Population* (Malthus), 239n32
eukaryotes, 192
evolution, 5, 49, 191–192
'Eve' mitochondria, 57–59, 65
*Everything Is Predictable* (Chivers), 259n6
extinction, 13–14, 63, 90, 158–163, 174–175, 206–207
extreme weather events, 139

Falkland Islands, 229
famine, 95–98
Fertile Crescent, 88–89
fertility, contraception and education, 17–18, 118–126, 127–130, 154, 186–187; *see also* total fertility rate (TFR)
financial crisis of 2008, 130
fig wasps, 205–206
fire, discovery of, 210–211
*First Steps: How Upright Walking Made Us Human* (DeSilva), 238n14
fitness, genetic, 163–164, 166
flooding, 140, 141
food insecurity, 96–98, 130–131, 189–190, 200–202
foramen magnum, 36, 41
founder effect, 101–105, 113, 165–166
*From Arabia to the Pacific* (Dennell), 74

# INDEX

*The Future of Geography* (Marshall), 268n28
fungi (pathogenic), 106, 107

Gandhi, Indira, 118
Gaucher disease, 102
genetics: chimpanzees, 12; genetic code, 57; genetic diversity (variation), 11, 12, 110; genetic engineering/modification, 196, 199, 200–201; genetic fitness, 163-164, 166; Gene Revolution, 22; *Homo sapiens*, 12, 163
giant deer, 90
Gibbon, Edward, 9, 10, 14, 82, 237n11
*Gigantopithecus*, 241n11
gluteus maximus, 41
Gott, J. Richard, 155-156, 172, 233
*Graecopithecus*, 33-34
Great Leap Forward, 16, 96, 123
Green Revolution, 22-23, 178-181, 184-186, 188-189, 195
gross domestic product (GDP), 126-127, 132, 151, 152, 184
the Gulf, 146

habitat destruction, 158-159
Habsburgs, 52-55
the Hajj, 146-147
Happisburgh, 47-48, 62, 138
Harper, Kyle, 107, 108-109
head louse, 79
heat, 143-145, 148

*The History of the Decline and Fall of the Roman Empire* (Gibbon) 9, 82
hobbits, 8, 10–11, 49–50, 51, 149
hominins, 10–11, 44–47
*Homo* (genus): appearance in Africa, 45
*Homo antecessor*, 24, 47, 50, 62, 66, 138, 149
*Homo erectus*: background, 11–12, 19, 21–22, 46–51, 157; discovers fire, 24, 206; migrates 64, 76–77, 149
*Homo floresiensis*, 8, 14, 49, 50, 80, 81
*Homo habilis*, 46
*Homo heidelbergensis*, 47, 49, 149
*Homo longi*, 49
*Homo luzonensis*, 8, 14, 49, 50, 80
*Homo naledi*, 52, 56, 62
*Homo rhodesiensis*, 51, 56, 62, 67
*Homo sapiens*: defined, 8–10; agriculture and 88–89, 204–205; causes extinction of other species, 13, 14, 56, 80, 82, 175, 182; bipedality, 41, 42; Earth system processes and, 182–184, 185; ecological niche of, 209, 210, 230; evolutionary diversification of, 21; extinction of, 6, 7, 20, 26, 59, 104, 105, 152, 157–158, 162, 164, 167, 171, 234; genetic diversity/homogeneity, 12, 63,

274

# INDEX

67, 103, 109, 113, 163, 164; interbreeds, 66, 167; migrates, 19, 64, 66, 74–79, 104, 149; modifies landscapes, 13, 203–209; origin of, 51, 56, 60, 61, 105, 156, 245; overexploits resources, 18, 131, 151, 183, 189–190; rarity of, 70, 71; sequesters products of photosynthesis, 23, 159, 189
*How Life Works* (Ball), 246
human(s): ageing, 125–126; definition of word, 236; migration of, 19, 149–150; human sperm count, 18, 132–134, 151–153; population expansion, decline, 15, 16–18, 20, 80–81, 89, 114, 120, 125, 129, 150; unknown species of, 51, 68–69
Human Genome Project, 60
Human Immunodeficiency Virus (HIV-1), 59, 109
humidity, 145–146
hurricanes, 142

ice age(s), 35, 67–68, 74–75, 86, 139, 166, 206–207
Iho Eleru, 51, 68, 69
immigration, 122, 124, 127, 128
immunity, 110
inbreeding, 16, 20, 52–55, 65, 101, 164, 166–167
Indonesia, 143
Industrial Revolution, 95–96
influenza, 93, 108–109, 111, 112

Institute of Chartered Accountants of England and Wales, 130
interbreeding, 61, 65, 66
intermediate disturbance hypothesis, 208
International Corn and Wheat Improvement Center (CIMMYT), 179
International Monetary Fund, 97–98, 131
International Rice Research Institute (IRRI), 179
International Space Station, 216, 219–220
in-vitro fertilization (IVF), 134
Irish Potato Famine, 15, 96
Ishango, 67–69
isostatic rebound, 139
Israel, 124

Jakarta, 143
Japan, 17, 122–123
Jews, 16, 102, 103, 124
Jurassic Period, 1
Justinian Plague, 112

Kalahari Desert, 104
Karenina Principle, 7, 11–12, 26, 162
KhoeSan people, 69
*Khoratpithecus*, 35
Kubrick, Stanley, 212
kuru, 107

lactose intolerance, 94
Lake Turkana, 63
Lascaux, 78
Leeuwenhoek, Antonie van, 181

# INDEX

Levine, Hagai, 132–133
lice, 79, 211
life expectancy, 125
light-harvesting complexes (LHCs), 193–194, 196–197
Lucy (specimen of *Australopithecus afarensis*), 240n4
Lukenya Hill, 68
Lunar Gateway, 220, 222

Makgadikgadi, 104, 105
Malthus, Thomas, 239n32
mammoths, 2, 79, 80, 90, 206
Mann, Charles C., 117–118
Maplecroft (risk assessment consultancy), 142–143
Mars exploration, 221–222
Marshall, Tim, 268n28
mass extinction, 174–175
Mediterranean, 145
megafauna, 13–14, 90, 206
metabolic syndrome, 94
migration, 149, 207, 231
milk, 93, 94
*Mir* (space station), 219
mitochondria, 57–60, 65
monsoon, 144, 209
moon, 220–221
Mount Carmel, 64
Mount Toba, 77
Murray, Christopher J. L., 121

Natufians, 88
natural selection, 5, 66, 102
*Nature* (journal), 60, 158, 183
*Nature Climate Change* (journal), 147
Nauru, 95

Neanderthals: discover fire, 210–211; DNA of, 20, 65, 69, 76–77, 79–80, 149; genetic diversity, fitness, 165–167; extinction of, 14, 20–21, 48–49, 56–57, 80–81, 164, 165; interbreeds, 64–66, 80, 167; origins of, 8, 62, 157, 237n8
*Neanderthal Man: In Search of Lost Genomes* (Svante), 237n9
New Orleans, 142
New York, 141–142
niche construction, 205–206, 209–210, 230
Nigeria, 124–125, 135
*Nomad Century* (Vince), 149

obesity, 94
*Oreopithecus*, 34
*Orrorin*, 36–37
orthogenesis, 4, 5
Ostrom, John H., 4
'Out of Africa' hypothesis, 60–61

Pääbo, Svante, 65, 237n9
*Paranthropus*, 45
parasites, 105
Permian Period, 174–175
photorespiration, 194–195, 196
photosynthesis, 23, 189–191, 195–199, 201, 218, 220–221
Photosystem I, II, 193–194, 197
photovoltaic cells, 199, 200
*Phytophthora infestans* (potato blight), 96

Pinnacle Point, 72
plagues, 93, 106–109, 112, 150
*Plagues Upon the Earth* (Harper) 107
planetary boundaries, 181–182
plesiosaurs, 2–3
Pliocene epoch, 44
population, *see* human(s)
*The Population Bomb* (Ehrlich) 17, 22–23, 115–120, 129, 153, 173, 177–178
porphyria, 16, 100–101
primates, 24, 55–56
prions, 106–107
pro-natalist policies, 128
protein, edible, 201–202
protists, 106–108
pterosaurs, 2

rainforest, 21, 77
Rapa Nui, 160–162
Raup, David, 174
Red Queen hypothesis, 5
resource extraction, 151–152
respiratory syncytial virus (RSV), 108
ribulose bisphosphate, 194
ribulose bisphosphate carboxylase (RuBisCo), 194, 195, 197, 198
Rift Valley, 44
River Thames, 138
Rockström, Johan, 181–182
*Roe v. Wade*, 117
Rutherford, Adam, 246

*Sahelanthropus*, 35–36
*Salyut-1*, 219
*Scientific American*, 234–235
seduheptulose-1 7-bisphosphatase (SBPase), 196
sex, 110
SARS-CoV-19, COVID-19, 16, 93, 109, 112, 130
Sheringham, 139
Shipden, 137–138
*Silent Spring* (Carson), 117
*Sivapithecus*, 35
Skakkebæk, Niels, 133
*Skylab*, 219
smallpox, 106, 110, 111–112
snow monkey, 24
space: colonization/exploration of, 21, 26–27, 119, 172–174, 183, 211–212, 216, 219–221; habitats in, 219, 222–226, 233; human health in, 220, 226–227; legal and governmental issues, 228–229
Space Shuttle, 219
sperm count, 18, 132–134, 151–153
*SPQR: A History of Ancient Rome* (Beard), 238n11
Still Bay technocomplex, 72, 71
Stringer, Chris, 60, 61
Stuart, Anthony J., 238n18
sweating, 145–146
syphilis, 112

technology, 73–74, 212–213
temperature, 141, 144–146, 210
Tiangong space station, 219
Tibetans, 52, 66

## INDEX

total fertility rate (TFR), 122–125, 127–128, 129, 130, 154–155, 186
'transported landscapes', 208
Triassic Period, 1
tuberculosis (TB), 93, 102, 106, 109

Ukraine, 96–97, 130
United Kingdom, 105, 117, 124, 144, 205, 229
United Nations, 121, 221, 228, 229
United States of America, 123–124, 142, 195, 231

vaccinations, 110–111, 151
*Vanished Giants* (Stuart), 238n18
Van Valen, Leigh, 5
vertebrates, 39
Vince, Gaia, 149
viruses, 106–108; *see also* COVID-19 (SARS-CoV-19)

Wales, 143–144
War of the Spanish Succession, 53–54
Weinersmith, Kelly & Zach, 226, 268n28
wet-bulb temperature, 145–147
Wilson, Allan, 57–58, 60–61, 65, 67
women, empowering, 17, 119, 125, 129–130, 150–151, 173, 187
woolly mammoth, rhinoceros, 90
World Bank, 131
World Economic Forum, 132
World Health Organization (WHO), 135
worms (parasitic), 92, 105–108

Y-chromosome, 60
Yemen, 147
*You Will Go to the Moon*, 27, 230

zoonoses, 108–109

## *About the Author*

John Gilbey

**Henry Gee** is a senior editor at *Nature* and the author of several books, including *A (Very) Short History of Life on Earth*, which won the Royal Society Science Book Prize in 2022. He has appeared on BBC television and radio and NPR's *All Things Considered*, and he has written for *The Guardian*, *The Times* (London) and *BBC Science Focus*. He lives in Cromer, Norfolk, England, with his family and numerous pets.